A SHORT
HISTORY OF
RUSSIA

ALSO BY MARK GALEOTTI

Armies of Russia's War in Ukraine

We Need to Talk About Putin: Why the West Gets Him Wrong, and How to Get Him Right

Russian Political War: Moving Beyond the Hybrid

Kulikovo 1380: The Battle That Made Russia

The Vory: Russia's Super Mafia

The Modern Russian Army 1992–2016

Spetsnaz: Russia's Special Forces

Russia's Wars in Chechnya 1994–2009

Russian Security and Paramilitary Forces since 1991

Paths of Wickedness and Crime

Gorbachev and His Revolution

The Age of Anxiety: Security and Politics in Soviet and Post-Soviet Russia

Afghanistan: The Soviet Union's Last War

The Kremlin's Agenda: The New Russia and Its Armed Forces

A SHORT HISTORY OF
RUSSIA

HOW THE WORLD'S LARGEST
COUNTRY INVENTED ITSELF,
FROM THE PAGANS TO PUTIN

MARK GALEOTTI

HANOVER
SQUARE
PRESS

HANOVER
SQUARE
PRESS™

Recycling programs
for this product may
not exist in your area.

ISBN-13:978-1-335-14570-3

A Short History of Russia

First published in 2020 in the United Kingdom by Ebury Press, part of the Penguin Random House group of companies.

This edition published by arrangement with Harlequin Books S.A.

Library of Congress Cataloging-in-Publication Data has been applied for.

Hanover Square Press
22 Adelaide St. West, 40th Floor
Toronto, Ontario M5H 4E3, Canada
HanoverSqPress.com
BookClubbish.com

Printed in U.S.A.

"Russia is a country with a certain future;
it is only its past that is unpredictable."

- Soviet joke

CONTENTS

INTRODUCTION

The oldest book in Russia does not speak with one voice. It roars and whimpers, mutters and moans, laughs and whispers, prays and brays, in progressively quieter tones. In July 2000, archaeologists excavating one of the oldest quarters of one of the oldest cities in Russia—Novgorod, once known as Lord Novgorod or Novgorod the Great—discovered three wooden boards, coated with wax, that together once were bound together as a book. According to carbon dating and other assessments, they were from somewhere between 988 and 1030 AD. Scratched onto the wax tablets are two psalms. This is a palimpsest, though, a document that has been used and reused, time after time, over decades, and yet on which the earlier writings are still just about visible. Painstaking work by the Russian linguist Andrei Zaliznyak uncovered a bewildering array of different writings once etched into the wax, thousands of them, from the "Spiritual Instruction for the Son from a Father and a Mother" to the beginning of the Apocalypse of John, a list of the Church Slavonic alphabet, even a treatise "On Virginity."

This is wholly fitting.

Palimpsest People

Russia is a country with no natural borders, no single tribe or people, no true central identity. Its very scale astounds—it stretches across 11 time zones, from the European fortress-region of Kaliningrad, now cut off from the rest of the Motherland, all the way to the Bering Strait, just 82 kilometers (51 miles) from Alaska. Combined with the inaccessibility of many of its regions and the scattered nature of its population, this helps explain why maintaining central control has been such a challenge, and why losing that grip on the country such a terror for its rulers. I once met a (retired) KGB officer who admitted that "We always thought it was all or nothing: either we held the country in a tight fist, or else it would all fall apart." I suspect his predecessors, from tsarist officers to early medieval princes, had much the same concerns—and Putin's officials, even with all the advances of modern communications, certainly do today.

Its position at the crossroads of Europe and Asia also means that Russia is everyone's perennial "other," with

Europeans considering it Asian and vice versa. Its history has been shaped from without. It has been invaded by outsiders, from Vikings to Mongols, crusading Teutonic orders to the Poles, Napoleon's French to Hitler's Germans. Even when not physically beset, it has been shaped by external cultural forces, forever looking beyond its borders for everything from cultural capital to technological innovation. It has also responded to its lack of clear frontiers by a steady process of expansion, bringing new ethnic, cultural and religious identities into the mix.

Russians are thus themselves a palimpsest people, citizens of a patchwork nation that more than most shows these external influences in every aspect of life. Their language is testament to this. A railway station is called a *vokzal*, for example, after London's Vauxhall station, the result of an unfortunate translation mishap when an awestruck Russian delegation was visiting nineteenth-century England. At the time, though, the Russian elite spoke French, so they will nonetheless load their *bagazh* onto their *kushet* sleeper car. In Odessa, to the south, streets were named in Italian because that was the common trading language of the Black Sea; in Birobidzhan on the Chinese border, by contrast, the local language is still to this day Yiddish, from when Stalin sought to encourage Soviet Jews to resettle there in the 1930s. In the fortified kremlin of Kazan, there is both an Orthodox cathedral and a

Muslim mosque, while shamans bless oil pipelines in the far north.

Of course, all peoples are compounds of different faiths, cultures and identities to greater or lesser extent. In an age when the curry is Britain's favorite dish, when the *Académie française* continues to fight its losing battle to keep French free of foreign words, and when more than one in eight US citizens are foreign-born, this is surely a given. But three things are striking about Russia's experience. The first is the sheer depth and variety of this dynamic, magpie appropriation of outside influences. The second is the specific ways in which successive layers have built upon each other to create this particular country and culture. All nations may be compounds, but the ingredients and the ways they mix vary widely. The third is the Russian response to this process.

Conscious—often self-conscious—of this fluid, crossbred identity, the Russians responded by generating a series of national myths to deny or celebrate it. Indeed, the very foundation of what we now call Russia has become shrouded in such a national "Just So" story, as I will discuss in chapter one, with conquest by Viking outsiders rewritten as the conquered themselves inviting the invaders in. Since then there has been a stream of such legends, from the way Moscow became both Christian and the "Third Rome," cradle of true Christendom (after the first fell to the barbarians and

the "Second Rome," Byzantium, to Islam), to today's
attempts by the Kremlin to present Russia as the bastion of traditional social values and a bulwark against
an American-dominated world order.

Back to the Future

The Mongols conquered Russia in the thirteenth
century and when their power ebbed, their most efficient quisling allies, the princes of Moscow, reinvented
themselves as their nation's greatest champions. Time
and again, Russia's rulers would edit the past in the
hope of building the future they wanted, typically by
scavenging the cultural or political myths and symbols
they needed. The tsars co-opted the symbols of glorious Byzantium but, in this case, the double-headed
eagle of empire looked west, as well as south. Over
the centuries, Russia's complex relationship with the
West would become increasingly crucial. Sometimes
this meant adopting ideas and adapting values to fit,
from Tsar Peter the Great ordering Russians to shave
their chins in European style (or pay a special "beard
tax") to the Soviets building a whole society on their
notion of an ideology that Karl Marx had envisaged
applied to Germany and Britain. Sometimes, it meant
a self-conscious determination to reject Western influences, even by redefining the past, such as by ignoring
all the archaeological evidence that the origins of this

land came with Viking invaders. Yet it never meant ignoring the West.

Today, hoping to be able to find a narrative allowing them to pick the aspects of the West they like—
iPhones and London penthouses without progressive
income tax and the rule of law—a new elite has again
begun trying to define themselves and their country
as suits their convenience. Not always successfully and
not at everyone else's convenience, though: over time
they came to question not so much their place in the
world as the way that world was treating them. This is
at the heart of the process that led to the rise of Vladimir Putin, and his evolution from an essentially open-
minded pragmatist to the nationalist war leader who
annexed Crimea in 2014 and stirred up an undeclared
conflict in southeastern Ukraine. This has become a
country in which reimagining history has become not
just a national pastime but an industry. Exhibitions
chart the lineages of modern policy back to the medieval era, as if in a single, unbroken evolution. The
shelves of bookshops groan with revisionist histories
and school textbooks are being rewritten in line with
new orthodoxies. Statues of Lenin rub shoulders with
those of tsars and saints, as if there are no contradictions in the visions of Russia they embodied.

The basic theme of this book, then, is to explore the
history of this fascinating, bizarre, glorious, desperate, exasperating, bloody and heroic country, especially

through two, intertwined issues: the way successive influences from beyond its borders have shaped Russia, the palimpsest nation, and the ways Russians came to terms with this through a series of convenient cultural constructions, writing and rewriting their pasts to understand their presents and try to influence their futures. And how, in turn, this came to affect not just their constant nation-building project but also their relations with the world. It is unapologetically written not for the specialist but for anyone who is interested in the backstory of a country that can at once be written off as a shambolic relic of an old empire, and at the same time be painted as an existential threat to the West.

In condensing a thousand years of eventful and often gory history into this short book, I have inevitably painted with a broad brush. At the end of each chapter I provide a guide to further reading that is much more scholarly and which can help restore the balance. Nonetheless, the aim of this book is not to pretend to be a comprehensive treatment of every detail, it is instead to explore the periodic rises and falls of this extraordinary nation, and how the Russians themselves have understood, explained, mythologized and rewritten this story.

Further reading: For the broad sweep of Russia's thousand years, there are many fine books I could recommend for particular elegance of approach or quirkiness of style, but let me note a few. Geoffrey Hosking's *Russian History: A Very Short Introduction*

(Oxford University Press, 2012) is exactly what it claims to be. A journalist's rather than a scholar's book, Martin Sixsmith's *Russia: A 1,000-Year Chronicle of the Wild East* (BBC, 2012) is a lively and readable overview. *Natasha's Dance: A Cultural History of Russia* (Penguin, 2003) by Orlando Figes focuses more on the past two centuries, but is nonetheless a tour de force. If a picture is worth a thousand words, a map is worth at least that, and Martin Gilbert's *Routledge Atlas of Russian History* (Routledge, 2007) is a very handy compilation. Histories are also written in brick and stone, though, and Catherine Merridale's brilliant *Red Fortress: The Secret Heart of Russia's History* (Penguin, 2014) takes the Moscow Kremlin as itself a character in Russia's story.

A Note on Language

There are different ways of transliterating from Russian. I have chosen to render words in Russian as best they sound, except when there are forms that by now are too established to be worth challenging: Gorbachev rather than the more phonetically accurate Gorbachov, for example. Language is intrinsically political, as how we talk about something conditions how we think about it, and this has become especially evident in post-Soviet times as states assert their independence from the metropolis, and with it their linguistic autonomy. This is a particular issue for Ukraine: nowadays, its capital is rendered as Kyiv. However, I still use the term Kiev for the pre-1991 city, not in any way to challenge Ukraine's claim to statehood, but to reflect the extent to which it was once part of a wider Slavic and then Russian political order. I also turn Russian words into plurals by adding -s rather than the correct -y or -i. My apologies to the purists.

1

"LET US SEEK A PRINCE WHO MAY RULE OVER US"

Timeline

862?	Arrival of Ryurik, birth of the new Rus' nation
882	Oleg takes Kiev and moves his capital there from Novgorod
980	Vladimir the Great becomes Grand Prince of Kiev
988	Vladimir decrees conversion to Orthodox Christianity
1015	Vladimir's death triggers dynastic struggles
1036	Yaroslav the Wise controls all the Rus' lands
1054	Yaroslav's death triggers dynastic struggles
1097	Lyubech conference
1113	Vladimir Monomakh becomes Grand Prince at the request of the people of Kiev

PUBLIC DOMAIN

Viktor Vasnetsov, Arrival of Ryurik to Ladoga *(1909)*

Viktor Vasnetsov's depiction of Prince Ryurik's arrival on the shores of Lake Ladoga is a classic of its kind. The twelfth-century *Primary Chronicle*, our best single source on the era, speaks of the skirmishes that the scattered Slavic tribes of what was to become Russia fought against the "Varangians"—their name for the Vikings of Scandinavia—to drive them out of their lands. But when the Chud and the Merias, the Radimiches and the Kriviches, and all the other myriad clans and tribes tried to govern themselves, the result was just more wars. Unable to agree on precedence and protocol, territory and turf, they turned again to the Varangians and sought a prince: "Our land is great and rich, but there is no order in it. Come rule and reign over us."

They got Ryurik (reigned 862?–79), the man whose descendants would form the Ryurikid dynasty that ruled Russia all the way through to the seventeenth century. Vasnetsov shows him landing on the shores of Lake Ladoga from his distinctive dragon-prowed Viking longship, along with his brothers and retinue, ax in hand to emphasize that he is a warrior-prince. There, he is being greeted by a delegation of his new subjects with both tribute and, quite literally, open arms.

The painting is characteristically detailed and evocative. It is faithful to the tale, down to the conical helmets of the Vikings and the traditional embroidery of the Slavs' clothing. It is artfully symbolic, with the tribute bridging the new ruler and his new subjects. It is also quite, quite wrong.

The Arrival of the Ryurikids

There was a Ryurik, possibly one Rorik of Dorestad, an ambitious Danish upstart whose raids so enraged Louis the Pious, King of the Franks, that he was banished in 860. This conveniently coincides with the date of Ryurik's arrival—generally set somewhere between 860 and 862—and his disappearance from Western chronicles. Scandinavian raider-traders had long known of the lands of the Slavs, not least in their quest to find new trading routes to Miklagarðr, "Great City"—the Eastern Roman capital of Byzantium, today's Istanbul—far to the south. The Byzantine emperor's elite Varangian Guard was drawn from Scandinavian mercenaries, after all.

Thus, when Rorik of Dorestad found himself dispossessed at home, why not carve himself a new principality in these territories? First, he established a fort at Ladoga, where he and his men had landed, and soon he would take over a trading post inland and establish a base. This he called Holmgarðr, although it would come to be known as Novgorod ("New City"), one of the great centers of old Russia. However, the evidence that he was invited in seems, alas, distinctly lacking.

Ryurik's adventure was just part of a wider drift of Scandinavians southward and eastward. They were sometimes traders but more often invaders in hostile lands, in savage competition with each other, not just the locals. The tenth-century Arab chronicler Ibn Rusta would later say, admittedly rather fancifully, that they so mistrusted each other and the peoples around them that a man could not go outdoors to relieve himself without being accompanied by three armed companions to guard him. Despite the dangers, though, the appeal of these lands was irresistible.

To the south and the east were the rolling plains of the steppe, the domain of various Turkic tribes, nomads and once-nomads, such as the Bulgars and the Khazars. They demanded tribute of neighboring Slav tribes, such as the Polyane ("plains people") who lived around the southern city of Kiev, but did not conquer or settle in their lands. Farther to the southwest was Constantinople, known to the Slavs as Tsargrad,

"Emperor-City." Its trading stations reached as far as the Black Sea, but it lacked the will, armies or interest to venture north. To the west were the Magyars and Western Slav peoples such as the Bohemians, in the process of creating their own, settled nations, increasingly dominated by the Germans.

In short, this was a land of many tribes and small settlements—the Scandinavians called it Gardariki, "land of towers"—but no kings. Broad, fast rivers, notably the Dvina and the Dnieper, the Volga and the Don, were virtual watery highways, crucial routes for raiding and trading by Varangians whose shallow-beamed boats would travel deep along them, and be carried or dragged the relatively short distances between. One could, for example, sail along the Neva from the Gulf of Finland to Lake Ladoga, as Ryurik did, then make for the source of the Volga, the longest river in Europe. With just a 5–10-kilometer (3–6-mile) stretch of portage—carrying boats overland—the travelers could then sail all the way south to the Caspian Sea. In these lands there was timber and amber, furs and honey, and the most lucrative commodity of all: slaves. More importantly, there were trade routes to Constantinople and directly to "Serkland"—Land of Silk—as the Muslim territories of the east were known. Scandinavians had extracted tribute in the forms of goods and silver from the northwestern tribes until the risings in 860 forced them from their timber-walled forts and back home, but it was hard to see why they would stay away.

Indeed, at around the time Ryurik was settling in Novgorod, two other Viking adventurers, Askold and Dir, had taken their men and seized the southwestern Slavic city of Kiev, using it as the base for an ambitious, if unsuccessful, raid on Constantinople. Others had already tried this, with adventurers from Scandinavia plundering the southern Black Sea coasts perhaps half a century before. The Slavs called these Varangian conquerors the Rus' (likely from the Finnish *Ruotsi*, their name for the Swedes) and so the lands of the Rus' were born.

Ryurik was succeeded by Oleg (r. 879–912), his war chief and the regent for his young son Igor. Oleg proved as efficient as he was ruthless, capturing and killing both Askold and Dir, and taking Kiev in 882. He moved his capital there from chilly northern Novgorod and this would remain the dominant city of the Rus' for centuries. When Igor (r. 912–45) replaced him as prince of Kiev in 912, the Ryurikid dynasty was truly born. Over time, the Scandinavian Rus' and their Slav and other subjects would intermarry, and their cultures intermingle. In some ways this was made easier by considerable overlaps in their pagan beliefs: the Slavs' Perun the Thunderer was very similar to the Scandinavians' Thor, for example. This way, in the wooden-stockaded towns and the little villages along the main river routes, as much forts as trading stations, a new nation was emerging.

© HELEN STIRLING

Conquest, trade, settlement and alliance would com-
bine to see Kiev's power grow. Raids on Constantinople and its lands would often be repulsed, but Kiev
also secured treaties in 907 and 911 that saw the greatest
city-state in the world treating upstart Kiev as, if not
an equal, nonetheless a power worthy of respect. Slav
tribes such as the Severiane and the Derevliane were
brought under Kiev's control, although not without
cost. (Igor would be killed by the latter, and bloodily
avenged by his widow, Olga.)

The Kievans were not without challengers, though.
They were conquerors, pirates and traders not just because of greed, but also need. A new nomad power
was rising in the south, the Pechenegs, and from 915
the *Primary Chronicle* details their increasing attacks,
especially on the rapids on the Dnieper, the river that
had become central to the trade and thus prosperity
of the Rus', and whose valley the Pechenegs treated
as their summer grazing and hunting grounds. Nine
granite ridges stretched across it, far southeast of Kiev.
In spring, when melting snows swelled the rivers and
burst their banks, these boat-breaking dams were sub-
merged, but at other times they formed barriers forcing travelers to pull their boats from the water and
drag them overland. At those times, the Kievans were
especially vulnerable to the Pechenegs, and Prince
Sviatoslav (r. 945–72) himself was killed trying to repel
an attack at the rapids, his skull ending up as a nomad

drinking cup. Just as the Kievans ran their own protection rackets among the Slav tribes and, when they could get away with it, neighboring peoples, they were periodically forced to pay off the Pechenegs.

Sviatoslav had been a warrior-prince, confident to the point of arrogance; his eldest son and short-lived successor, Yaropolk, appears to have been insecure to the point of fratricide. He killed his brother Oleg (who, in fairness, might have struck first) and forced his other brother, Vladimir, from his bastion of Novgorod. However, Vladimir would return in 980 with an army of Varangian mercenaries, kill Yaropolk and take the crown for himself. He would then go on to change the shape of Russian history.

Vladimir the Great

Vladimir (r. 980–1015) would prove an empire-builder. Whereas Sviatoslav was the classic Varangian prince, a hard-fighting warrior-raider who himself took an oar when sailing down to Tsargrad in the hope of plunder, Vladimir was a planner and a politician, and eager to take the Rus' beyond their Viking roots. He expanded the territories under his rule, battling the Pechenegs, conquering tribes, seizing towns and harrowing the Volga Bulgars. He had defenses thrown up around Kiev, with the mighty Snake Ramparts—which, when completed in the eleventh century, would stretch for a hundred kilometers—securing it from the south. New cities were founded at Belgorod and Pereyaslavl, along with fortified harbors along the Dnieper. A chain of forts was built to keep the Pechenegs at bay, with traditional wooden walls now reinforced with unfired bricks thanks to Greek builders from Tsargrad.

The reason why new techniques and technologies were now being imported from Constantinople was his fateful decision to convert to Christianity—and force the Rus' lords and subjects to follow suit. There was little early evidence of such inclination on Vladimir's part. He had previously ordered a pagan temple to be built on one of Kiev's hills, with great wooden idols looking down on the city, and seems to have shrugged off periodic mob violence against Christians. In 988, though, Vladimir ordered those idols cast down, and the population of Kiev herded at virtual spearpoint into the river Dnieper for forced baptism. (Although for centuries, Christianity and paganism would coexist, as the former only slowly displaced the latter in reality.) Faith and state power would begin the close alliance that defines Russia to the present day.

Why did Vladimir do this? The apocryphal tale is that he sent envoys out to assess the appeal of the main faiths dominant at the time. Judaism was rejected because he believed that for the Jews to have been expelled from their homeland proved that God was not on their side. Roman Catholicism was rejected because no Grand Prince of Kiev could submit himself to the authority of the Pope. Islam was rejected because of its prohibition of alcohol, with Vladimir allegedly noting that "drinking is the joy of all Rus'. We cannot exist without that pleasure." (Some stereotypes have a long pedigree, it seems.) Instead, it was Byzantine Orthodox Christianity that won him over, as his emissaries rhapsodized about the Eucharist in the domed nave of the immense Hagia Sophia cathedral, where "we knew not whether we were in Heaven or on Earth, nor such beauty, and we know not how to tell of it … We only know that God dwells there among the people, and their service is fairer than the ceremonies of other nations."

Well, maybe. Again, it's a lovely story, but the truth is likely rather more complex and pragmatic. Orthodox Christianity had been spreading amongst the Rus', and especially the boyars, their lords and chieftains. The Cyrillic language that would in due course become standard across all Russia has its roots in Greek, modified to cater for the Slavic tongues by Saints Cyril and Methodius, ninth-century Byzantine missionaries. Vladimir's own grandmother Olga (r. 945–60) would go on to be baptized as a Christian, not that she demonstrated a particular penchant for turning the other cheek: she was, for example, infamous for burying and burning Derevlian emissaries alive to avenge her husband, Igor. Besides, the religious riots that had rocked Kiev had demonstrated the risks of trying to ignore the tensions between pagans and Christians. Byzantine Christianity did not require submission to a distant spiritual leader, and brought the promise of closer relations with Tsargrad. According to some sources, Vladimir was already making inroads into Byzantine territories, having seized Chersonesos on the Crimean Peninsula; according to other, mostly Arab writings, the Greeks were rocked by civil war, and Emperor Basil II was desperate for allies. Either way, Vladimir seized on Byzantine weakness to seek a dynastic alliance. His aim was marriage to Anna, the emperor's sister; the price was not just military support but also the adoption of Christianity, both for himself and his people.

The deal was struck, with Vladimir baptized at Chersonesos. Later, he would be sanctified as Holy Grand Prince Vladimir, Equal of the Apostles, but this seeming act of piety was actually a piece of ruthless statecraft. It reaffirmed his status as greatest of the Rus' and cemented ties with their most powerful neighbor and richest trading partner.

Vladimir, who had experienced exile before, might also have thought of this as lining up a nice bolt-hole for the future, just in case he suffered another such reversal. As it was, though, he ruled for almost another three decades. As Kiev's lands expanded, and mindful of the difficulty of ruling such a sprawling state, he appointed his sons as princes beneath him to various cities. Tribute continued to flow to Kiev, but in such a time, and in such a land where roads were few, the river routes largely confined to north–south travel, and the lands between cities thickly forested and sparsely settled, the hand of the Grand Prince inevitably rested lightly on his vassals. The Grand Prince of Kiev simply could not control the day-by-day governance of the cities. A prince had his armed retinue, his cronies and his favorites, his own interests and priorities. Unless he was on the border and needed help repelling foreign enemies, why did he need to heed Kiev?

The first to test this was Yaroslav the Wise, who stopped sending his father tribute in 1014. Vladimir began to muster forces to reassert his power, but he was already ailing and died the next year, before he had a chance to launch his punitive expedition. The result was a bloody spat between pretenders, and one which for the first time saw the involvement of another rising power of the time, the Poles. Yaroslav's elder brother Sviatopolk had already conspired against his father, possibly encouraged by his father-in-law, the Polish Count Bolesław. Over the next few years, Kiev would be taken first by Sviatopolk, then Yaroslav, as they struggled for dominance. Yaroslav hired Varangian mercenaries, Sviatopolk Pechenegs and Poles. Yaroslav won, but he had established a dangerous precedent of bloody familial disputes. His nephew Briacheslav of Polotsk began casting hungry eyes at the rich markets of Novgorod, while his formidable brother Mstislav of Chernigov and Tmutarakan to the south was making moves toward Kiev. It took until 1036 for all other rivals to be eliminated and

Yaroslav (r. 1036–54) finally to be Grand Prince of Kiev, Prince of Novgorod, and ruler of all the Rus'.

It was a triumph. As with all high points, though, there is then nowhere to go but down.

Fragmentation and Rotation

Hard-won, Yaroslav's reign would then be one of paradoxical successes. He took back the lands seized by Bolesław, conquered territories in what is now Estonia and routed a Pecheneg siege of Kiev. While a naval assault against Constantinople in 1043 was a failure, he managed nonetheless to secure a new treaty with Tsargrad and marry one of his sons, Vsevolod, to another Byzantine princess (of which they seemed to have an inexhaustible supply). There was peace, trade and good harvests. Silver flowed from Constantinople, the Arab world and northern Europe. The towns of Russia prospered, their markets growing, their wooden walls being pushed out as more settled within their defensive embrace. With the white-walled and golden-domed cathedral of Saint Sofia completed in Kiev, other cities began laying the foundations of similar physical tributes to their new faith.

All of this was obviously a sign of progress and a boon to Kiev, with more tribute potentially available to the Grand Prince. However, this also held within it the seeds of political fragmentation. The lands of the Rus' were essentially treated as a family patrimony. The Grand Prince would assign cities to his sons as princes, or to

trusted *posadniks*, governors. Princes might be moved between cities as need and opportunity dictated. Yaroslav
had originally gone from Rostov to Novgorod in 1010,
for example, when his brother Vyacheslav had died, and
his younger brother Boris had occupied the city he left.

A prince's legitimacy rested on that of the Grand
Prince, and his military forces were limited. He would
have his *druzhina*, his personal retinue, but at most this
might number a couple of hundred men, enough to col-
lect taxes and guard the prince, not a war-fighting force.
Beyond that, he could hire mercenaries from abroad—
which came with its own costs and risks—and raise levies
from his own town. His capacity to do this often depended on the time of year (was everyone busy bringing
in the harvest?) and how popular he was with his people.

There was no clearly accepted process of succession,
though, explaining the fratricidal turmoil that followed
Vladimir's death, and then that of Yaroslav in 1054.
Should the position of Grand Prince go to the eldest
son or the oldest brother? The second half of the eleventh century saw periodic civil wars and uncivil spats as
these issues were fought over, half resolved, only to be
reopened. In part, this was precisely because the cities,
and the principalities that cohered around them, were becoming more powerful and prosperous. They provided
the economic basis that allowed a prince more scope to
wage his private wars. They also began to acquire their
own voices, especially through the *veche*, a town assem-

bly. In theory, the *veche* was a place where all free male citizens could be heard, although in practice it tended to be an instrument of the rich and powerful. In Novgorod, which, as will be discussed in the next chapter, had become a Baltic-facing trading city, the *veche* had an especially powerful role, even making its own decision as to who should be *posadnik*. However, elsewhere there were also signs that the townsfolk could become a political force in their own right. In 1113, for example, the people of Kiev successfully petitioned Prince Vladimir Monomakh of Pereyaslavl (r. 1113–25) to become their Grand Prince even though, according to the consensus of the princes, the crown should have gone to Yaroslav of Volynia. Monomakh hesitated to buck the princely collective, until the Kievans warned, "Come, prince, to Kiev. If you do not come, know that much evil will be stirred," going on to threaten everything from pogroms against local Jews to attacks on his own sister-in-law.

Here was the irony: through the twelfth century, control of Kiev and the position of its Grand Prince became increasingly prized and perennially fought over. Time and again, a new Grand Prince would have to fight off rivals, as multiple dynastic bloodlines struggled to assert their might or their right. Yet meanwhile, the Rus' state itself became politically fragmented, devolving into a confederation of principalities. At different times, some became especially semidetached, while others were especially closely tied to Kiev. Ultimately, they were all part of one Rus' community; they looked to Kiev as not just a prize but also a center of their culture, faith and identity. But the princes did not necessarily see themselves as subjects of the Grand Prince, and were perfectly willing to make their own policies. This was virtually formalized at the 1097 summit that took place in Lyubech, where, to present a united front against nomad incursions, they agreed to end the old system whereby princes could be rotated between cities. Instead, positions would be inherited within bloodlines, and territories divided on inheritance. Thus was born Russian feudalism, even if its rules were periodically bent and broken.

So Kiev rose, at the crossroads of multiple civilizations and polities, and as the capital of a land built on trade and conquest. In many ways, by the beginning of the thirteenth century, though, its ambitions had outrun its influence. The city was rich and respected, the site of growing industries from glassblowing to jewelry, the heart of the Russian Orthodox Church and the dream of every ambitious prince. It was not in charge, though, and power was slipping imperceptibly through its fingers. The archaeological evidence suggests that despite sporadic dynastic conflicts, Russia was prospering. Kiev was periodically sacked, but it nonetheless always managed quickly to rebuild and rebound. The Novgorodians were busy driving new trade routes into northern Siberia and their city was host to a thriving community of Baltic traders. The principality of Vladimir-Suzdal was pushing into the Bulgar lands. Even the rise of a new nomad threat, the Cumans and the Kipchaks, whom the Russians together called Polovtsians, was manageable. They first appeared in 1055, and by 1061 were raiding Rus' lands. Although they managed to defeat an army led by Vladimir Monomakh in 1093, he would rally and drive them out. They would continue to raid, often in force, plundering Kiev's Monastery of the Caves in 1096, but they would not now pose an existential threat to the Russians, and some of their subject tribes even ended up serving Kiev.

What the Russians didn't know was that behind the Polovtsians, and in due course pushing them westward before them, was a new nomad challenge, and one orders of magnitude more dangerous. The Mongols were coming, and the divided, self-absorbed principalities of the Rus' had no idea what that would mean.

The Normanist Conquest

The bloody ins and outs of ancient Rus' politics may seem distant, of little relevance today. However, any history of Russia must start with this not just for reasons chronological, but also because one can draw a direct and often-bloody line between these times and the present day. The origin story, in which vulnerability is spun as agency, sets the tone, especially as this is not simply a story of weakness, but of embracing conquest and creating something new from it. So many of the fundamental Russian assumptions about the world and their place in it can be traced back to the times of Ryurik and Vladimir, Yaroslav and their successors.

First of all, there is the constant struggle between center and periphery, which is an inevitable challenge in a sprawling land, even today in the age of modern communications. The resulting pattern of power, of myriad principalities merging, dividing and competing, "gathered" to the center and then again lost, would become fundamental to Russia. Secondly, that Russia was forever doomed to be surrounded by mighty powers who at once threatened and yet also impressed, whose cultural, technological, military and economic strengths were to be both resisted and emulated. This is, one could argue, forever the fate of a country at the crossroads of Europe and Asia, north and south, to be at once a vital waystation on the great trade routes along which flow ideas as easily as wealth, and also the target of whichever empire is on the rise.

To the north of the Rus', the Varangians had power and purpose, providing a new ruling class but also a constant challenge. To the south were the Pechenegs, whose speed and savagery made them a threat the Russians could at best hold back, but never conclusively defeat. Farther south, Constantinople offered cultural capital and trading might, such that the best to which Kiev could aspire was to be the "Tsargrad of the Rus'."

To the west, new challengers such as the Germans and the Poles were rising, and already they were beginning not only to gnaw at Russia's borders but also to interfere in its dynastic politics.

That was history, but history is very much alive in Russia today. Vladimir Putin has affirmed that "we have become aware of the indivisibility and integrity of the thousand-year-long history of our country." Russians are voracious consumers of films and books about their country's past. There is a passionate reenactment movement, with medieval warriors brawling in simulated melees in front of wooden-walled fortresses, and gorgeously uniformed grenadiers and hussars clashing in make-believe Napoleonic battles. To an extent, this is a rediscovery of stories freed from the chilling grip of Soviet orthodoxy, but it is also something assiduously encouraged by a state eager to mobilize it to its own ends. George Orwell's dictum that "he who controls the past, controls the future" may be overstating the case, but the Kremlin is certainly willing to give it a try.

Vladimir Putin's worldview is of a Russia likewise beleaguered, although now China is perhaps the new Constantinople, at once the object of fear, jealousy, greed and a desperate need for alliance, against what he sees as a hostile, prescriptive and degenerate West, and a turbulent Islam to the south. He draws repeated historical parallels, such as in 2014, when he warned that the West's "infamous policy of containment, initiated in the

eighteenth, nineteenth and twentieth centuries, continues today. They are constantly trying to sweep us into a
corner because we have an independent position."

The new standard line, that when Russia is weakened
by internal division, then it is prey to outside forces, has
obvious advantages for a government keen to encourage unity and paint the opposition as unpatriotic. Indeed, Putin actively seeks to portray the outside world
as trying to work with and through separatists, anti-
government campaigners and other political enemies
precisely to create that kind of disunity and vulnerability: "they would gladly let Russia follow the Yugo-
slav scenario of disintegration and dismemberment."

History becomes a guidebook for geopolitics. Because the Grand Princes of Kiev could not count on the
support of the other princes, the Pechenegs could raid
the Dnieper trade routes. Only when Kiev was strong
would the Greeks treat them as equals and the borders
be safe. The dynastic struggles that followed Vladimir
the Great's death shattered that unity and allowed Varangians, Pechenegs and Poles to play a role in Russian politics. The result was that Kiev was plundered and western
towns were traded to the Poles. In the epic *The Tale of
Igor's Campaign*, describing a twelfth-century struggle
against the nomads, the narrator inveighs against both
sides in a civil war of the day: "Lower your banners now
and sheathe your tarnished swords ... In your seditious-
ness, you began to incite the pagans against the lands of

the Rus'. Violence from the lands of the Polovtsy came about because of your strife." Today, Putin's propaganda machine similarly calls on dissidents and protesters to set aside their grievances with the state in the name of Russian unity and thus security.

But Putin is simply following in a centuries-long tradition of presenting the world as a hostile place full of ravening predators ready to pounce on Russia if ever it lowers its guard. Now, the danger is from division, but it could just as easily be backwardness. Russia had to keep up with its neighbors—whatever the cost—because otherwise it faced "constant defeats," as Stalin forcefully put it in 1941, justifying his murderous industrialization and collectivization programs that killed and immiserated millions, in the name of this savage vision of history:

> She was beaten by the Mongol khans. She was beaten by the Turkish beys. She was beaten by the Swedish feudal lords. She was beaten by the Polish and Lithuanian gentry. She was beaten by the British and French capitalists. She was beaten by the Japanese barons. All beat her because of her backwardness, military backwardness, cultural backwardness, political backwardness, industrial backwardness, agricultural backwardness.

As we shall see, sometimes it was indeed economic or political weakness that allowed Russia to be humbled by foreign powers over its long and bloody history. Often, though, it was not—or at least, this was not the whole answer. But such tedious objectivity is irrelevant to those who would build grand historical narratives and use them to justify what were often state-building schemes of extravagant scale and brutality.

Perhaps one of the reasons why Russia's history is at once so vividly real and conveniently malleable today is precisely because of the passionate way it has been rewritten over the centuries. New myths are superimposed over old ones in the creation of this palimpsest identity, as the peoples of this land sought to come to terms with their lack of strength and common identity by creating shared mythologies that saw fate and frailty translated into pride and purpose. They were not conquered by the Rus', for example; they invited their new princes in. That was still too much for a new generation of Russian historians in the eighteenth century, who branded this as "Normanism" and claimed, instead, that the Slavs had not needed any Varangians. Instead, they formed their state themselves, and the name Rus' was derived from the name of an ancient tribe. Historically dubious but nationalistically gratifying, this briefly became the new orthodoxy until the tiresome demands of scholarly accuracy relegated it to the margins. In Soviet times, though, the notion that their country could have Germanic origins was unacceptable, and a rejection of "Normanism" became for a while state dogma.

Vladimir's seizure of Chersonesos likewise became part of Moscow's rationale for its annexation of Crimea in 2014, on the grounds that it makes the peninsula the cradle of Russian Orthodoxy. Meanwhile, even as their troops battle over the Donbas region, the historians of Moscow and Kyiv battle over who can claim Vladimir the Great as their own: as Grand Prince of Kiev, does that mean that the spiritual ancestor of modern Russia is actually a Ukrainian, or does his Ryurikid pedigree prove that Ukraine is truly just a semidetached part of Russia? Ancient history, national myths and modern wars can be closer than we might like to believe, and nowhere more so than in the lands of the Rus'.

Further reading: There is, sadly, something of a dearth of readable studies of this fascinating time. Simon Franklin and Jonathan Shepard's *The Emergence of Rus 750–1200* (Longman, 1996) remains one of the foundational texts, although it is not the lightest of reads. That is doubly true of Pavel Dolukhanov's scholarly *The Early Slavs: Eastern Europe from the Initial Settlement to the Kievan Rus* (Routledge, 1996). The first half of Janet Martin's magisterial *Medieval Russia, 980–1584* (Cambridge, 2007) is a more accessible overview. Vladimir Volkoff's *Vladimir the Russian Viking* (Overlook, 2011) is a very readable biography of Vladimir the Great, but in fairness it should not be treated as a rock-solid historical account. I would finally note that David Nicolle's short *Armies of Medieval Russia, 750–1250* (Osprey, 1999) has more than just a useful summary of the wars and raids of the time, and pretty color pictures by Angus McBride.

2

"FOR OUR SINS, THERE CAME AN UNKNOWN TRIBE"

Timeline

Bas-relief of Prince Dmitry being blessed by Sergius of Radonezh (1849), Donskoi Monastery

In 1380, the chronicles tell, the armies of the Rus' principalities came together under Prince Dmitry of Moscow at a place called Kulikovo (Snipes' Field). There they faced the mighty hosts of the Golden Horde, the Mongol-Tatar overlords of Russia since their brutal conquest the century before. Outnumbered, the Russians nonetheless outsmarted and outfought their enemies, with the first blow being struck by the warrior-monk Peresvet. In so doing, they threw off the "Mongol Yoke" that had so oppressed them. "Prince Dmitry returned with a great victory, like Moses won against Amelek. And there was peace in the Russian land. And his enemies were put to shame," wrote a chronicle. The bas-relief pictured here, once in the great Cathedral of Christ the Saviour in Moscow, then moved to the Donskoi Monastery when Stalin had the building blown to rubble, shows a classic scene from the story. A pious Prince Dmitry kneels to receive the blessing of St. Sergius of Radonezh, one of the great figures of Russian Orthodoxy, surrounded by knights from across Russia. Behind Sergius stands Peresvet, about to be commended into the prince's service.

The aforementioned chronicle was *The Life and Death of Grand Prince Dmitry Ivanovich*, an overblown, hyperbole-heavy and fact-light eulogy commissioned and written shortly after his death in 1389. It was part of two, intertwined processes of mythmaking, as both Moscow's rulers and the church tried to place themselves at the heart of a Russian reemergence that was, itself, still very much a work in progress. Peresvet probably never existed. Dmitry's army was by no means made up of contingents from every Russian principality; some stayed away, some such as Ryazan actually joined the other side. While Dmitry did indeed win the battle, gaining as a result the soubriquet *Donskoi*, "of the Don," after the river on whose banks it was fought, this was not the decisive victory often claimed. Just two years later, an army would sack and burn Moscow and force Dmitry to reaffirm his allegiance to the Golden Horde. For another century, the Russians would still have to send caravans of silver in tribute to the Horde's capital in distant Sarai.

For most Russians, though, this was not such a big deal. Indeed, the term "Mongol Yoke" would not have been recognized by anyone at the time. The Mongols were savage in their conquest, but surprisingly even-handed in their rule, and for most Russians it made little difference whether the silver was heading to Kiev, Moscow or Sarai. It was really only later that this era was turned into such a pivotal moment in Russian history, but the power of this myth can hardly be overstated. It shapes modern political culture, attitudes to China and even liberals' laments about why their country isn't more European. Perhaps above all, the irony is that it was the Mongols who enabled the rise of what had previously been a petty little hunting-lodge village into the heart of the new Russia: Moscow.

The Three Brothers

A recurring theme in Russian folklore is that of the three brothers. There was Lech, Czech and Rus, the supposed founders of the three Slavic peoples: the Poles (Lechites), the Czechs and the Rus'. Most tales of three brothers, though, feature one who is strong and fair, one smart and adventurous, one—the youngest—either rotten to the core or else a holy fool.

Well, once upon a time, there were three cities, and in many ways each represented a different path that Russia could have taken. Kiev was the greatest, and in many ways the most traditionally feudal. Power was expressed through family lineage and the common belief that Kiev was the heart and soul of the Rus'. It would regularly be fought over as one prince or lineage tried to assert a claim, but all the pretenders essentially shared the same worldview. One aspect of this was a constant struggle between princes wanting to gather more lands to themselves, and the custom of appanage, of sharing inheritances between all of a lord's sons, constantly fragmenting holdings that had been painfully and bloodily assembled. Kiev was a princely city, and for all its monkish chronicles of exalted piety and humility, its boyar aristocracy rose because of their skill and good fortune in the traditional arts of making war, conducting subterfuge, securing dynastic alliances and extorting tribute.

Novgorod to the north was a trading city, one whose reach extended into the Baltic Sea and its wealthy and cosmopolitan ports. More power lay with rich magnates and a rough-and-ready kind of oligarchic democracy. The *veche*, or assembly of the city's freemen, had a real voice and the annually elected *posadnik*, mayor, was often a greater power than the prince. Tellingly, the city was often called "Lord Novgorod the Great" as if it was its own master, and the prince was seen more as its employee. According to the *Novgorod Chronicle*, for example, in 1136 the people of the city decided to throw out Prince Vsevolod, "and they made these his faults.

I. He has no care for the peasants. II. Why did he want to rule in Pereyaslavl? III. He fled and left the army behind" in a recent war. To the Novgorodians, a prince was their figurehead and war leader. If he was neglecting his people, obviously eager to be in another city and not willing to lead their forces in battle, then he wasn't up to the job. He was expelled, and though other princes would be more effective in dominating this self-willed city, Vsevolod was neither the first nor the last to be weighed and found wanting. In 1270, for example, one Prince Yaroslav found himself on the wrong side of the *veche* "and the men of Novgorod answered: 'prince, go away, we do not want you. Or else we shall come, all of Novgorod, to drive you out.'" He left.

So Novgorod had its own culture. Christianity took longer to take root there—as late as 1071 there were pagan riots—and the boyars who dominated the city's politics were as much commercial magnates as warriors. The city's markets and trading stations made it a crucial source of silver for all the Rus' and paid for the food it needed to have shipped up the Volga. They also meant Novgorod was almost as much a northern European city as a Russian one. It was deeply involved in Baltic politics, clashing with Sweden, fighting off the muscular Christian raids of crusading orders such as the Brothers of the Sword and the Teutonic Knights, backing its allies in Livonia, being part of the intellectual currents of northern Europe. In short, Novgorod was a mercantile city, and its boyars prospered through trade and exploration, in what could, with a degree of optimism and oversimplification, be seen as a little like a very early, wooden-walled and onion-domed answer to a Renaissance Italian city-state.

When both Kiev and Novgorod were at their height, though, the third and youngest brother, Moscow, was scarcely a township. The first reference to it was in 1147, when Yuri *Dolgoruky*, "Long-Armed Yuri," soon to be Grand Prince of Kiev, arranged a meeting there. Nonetheless, when the Mongols came, Kiev would be broken and Novgorod humbled, and it would be Moscow that would thrive. This was the city that would not only become master of all the Rus' but also impose its own political culture, a fusion of Russian tradition, Mongol practice and Muscovite pragmatism.

The Coming of the Mongols

Nomadic and semisettled peoples to the south and east had been a perennial problem for the Rus'. The Judeo-Turkish Khazar Khanate of the Black Sea steppe had challenged them for control over the Volga trade routes in the ninth century. The Pechenegs from Central Asia had been a threat in the tenth, but would find themselves penned behind Rus' forts to the west and assailed by new challengers from the east. These especially included the Cumans, also known as the Polovtsians, a serious problem in the eleventh and twelfth centuries. However, none were a truly existential threat to the Russians. Indeed, when not raiding the Rus', they were trading with them or serving as mercenaries in one or another of the constant dynastic struggles that seemed the closest thing Russia's princes had to a common hobby.

Although no one knew it at the time, though, far to the east a power was rising that would reshape Eurasia. In the late twelfth century, the Mongol warrior Temujin, later to be known as Genghis Khan, was uniting an alliance of nomad peoples. He began an era of conquest such as the world had never seen, and his successors would come to regard themselves as having a divine mandate to extend their rule across the world in the name of Blue Sky Tengri, the ruling deity of their shamanic faith. Settled peoples were conquered; other nomadic powers incorporated or destroyed. China, Central Asia, much of the Middle East, all would fall to this formidable steppe army, its savagery, its speed and also its capacity to wield diplomacy, disinformation and despair as effectively as bow and sword.

By the early thirteenth century, the Polovtsians, who had displaced the Pechenegs, found themselves in the same invidious position, facing a newer, greater, sharper-toothed nomad threat from the east. Polovtsian Köten Khan fled to the court of Prince Mstislav the Bold of Galich, his son-in-law, with a stark warning: "Terrible strangers have taken our country, and tomorrow they will take yours if you do not come and help us." As new accounts came of a Mongol army on the banks of the Dniester River, Mstislav marched to meet them. The Russians and their Cuman allies were led into a trap and crushed at Kalka River.

However, this had merely been the advance guard of the invading force of the military commander Jochi Khan, Genghis's oldest son, and so the Mongols failed to follow up on this victory. The Russians had no idea of the size and ferocity of the threat they faced, and the Mongols—also known as Tatars—seemed as mysterious as they had been murderous. As the *Novgorod Chronicle* put it, "for our sins, there came an unknown tribe. No one knew who they were, or their origin, faith, or tongue … Only one Russian warrior in ten lived through this battle." As no immediate invasion followed, the Rus' managed to persuade themselves that the Mongols, whoever they were, had been deterred by their plucky resistance, and somehow a crushing defeat came to be seen as a bold defiance.

Until 1236, that is, when Jochi's son Batu Khan led his main force westward. The last remnants of the Polovtsians were crushed, and then he turned to the Rus'. The next few years saw the Russians harrowed by a storm of fire and blade. City after city was taken: divided and unprepared, the Rus' could not stand against the invaders. Proud Kiev fought, and was sacked with such murderous savagery that it was said that only 2,000 of its

50,000 population survived. Six years later a papal envoy wrote of ruins strewn with "countless skulls and bones of dead men." Canny Novgorod learned the lesson, pre-emptively buying its survival with silver and surrender.

It seemed nothing could stop the Mongols' inexorable westward march of conquest. Batu's armies continued on to Hungary and Poland, when politics and wine did what no army had yet managed to do. The Great Khan, Batu's uncle Ogedei who had succeeded Genghis, had a notorious fondness for drink—when his courtiers tried to limit the number of cups he was brought a day, he simply had them made bigger—and he died in 1241 after a night-long binge. While Batu continued to triumph in the field, he now faced what was likely to be a sequence of sieges where the cost of victory might outweigh the value of the plunder, and in unusually warm, wet weather at that, bogging down the usual lightning-fast Mongol cavalry. Perhaps it was with a sense of relief that, when he heard of Ogedei's death, Batu returned to distant Karakorum, the Mongols' capital, to take part in the negotiations that would select the new Great Khan.

The Mongol armies withdrew from Central Europe—but Russia remained under their control, part of the territories of the Golden Horde, as the western portion of the sprawling Mongol holdings was called. In conventional Russian accounts, what followed was more than two centuries of Asiatic despotism, during which time the Russians groaned under the "Mongol Yoke" and were locked away from the rest of Europe. The truth, of course, is rather more complex.

The Mongol Yoke?

The Mongols were more about conquest than administration. The Golden Horde, which would become increasingly autonomous as the Mongol Empire was just too large to rule as a single unit, built a capital for itself at Sarai, southeast of Russia, close to where the lower Volga flows into the Caspian Sea. The Mongols were not interested in imposing their own ways or faith on their subject peoples (from the mid-thirteenth century, Islam became dominant among them). Instead, they expected order and tribute, submission and obedience, and to that end were happy to rely on subject princes who could provide them. While at first the Mongols appointed their own local governors, *baskaks*, they were soon withdrawn as Russian princes proved willing to work for the Golden Horde instead.

Most were already used to being at best big fishes in small ponds, owing allegiance and tax to another. In many ways it made little difference to them just who was at the top of the food chain. They would make the journey to Sarai in hope of receiving the *yarlyk*, or mandate, to rule their principalities in the khan's name. From time to time a city would rebel or fail to show the requisite deference to some traveling Mongol dignitary, and the result would be a bloody reprisal. More often, rival princes would seek to enlist the support of Sarai in their own private feuds and dynastic struggles.

On the whole, this was a time of religious tolerance (in 1267, the Great Khan explicitly brought the Russian Orthodox Church under his protection, exempting it from tax and military service) and burgeoning trade. It was a particular opportunity for the town of Moscow, and the Ryurikid dynasty which held it. Moscow had been sacked and burned like so many others during the initial invasion, but as it recovered, its princes proved to be the quickest and most effective at understanding the new rules of the game. They would become the Golden Horde's most enthusiastic, effective and ruthless proxies. Whether it was raising taxes or punishing rebels, the Muscovite princes would eagerly do what Sarai needed—and make sure they profited richly in the process.

Alexander Nevsky, who had been prince of Novgorod before the Mongol invasion, saving northwestern Russia from the Teutonic Knights (Catholic crusaders who considered the Russian Orthodox as heathens, no better than Muslims), had from the first supported conciliation toward the Golden Horde. In return, he was awarded the *yarlyk* of Grand Duke of Vladimir-Suzdal, which had supplanted Kiev as, in effect, the mark of being the first among the Russian princely titles. The Ryurikids tried, with considerable success, to keep this title in the family pretty much consistently thereafter, and with it the prestige and opportunities for enrichment it offered. Moscow was one of the cities in his patrimony, the least important, so when he died it went to his youngest son, the two-year-old Daniil.

Nevsky's successor, Yuri (r. 1303–25), spent two years schmoozing and politicking in Sarai and married Konchaka, sister of Uzbeg, Khan of the Golden Horde. Not least thanks to that inspired alliance, he was also made Grand Prince of Vladimir-Suzdal. However, Moscow was still locked in a struggle for dominance over the Russian principalities with Ryazan and, especially, Tver. On Yuri's death, Prince Ivan I (r. 1325–41) neatly killed two birds with one stone by volunteering to lead the suppression of a rising in neighboring Tver. He not only got the chance to lead a Mongol army against his rival, he was also granted the usual prize of being made Grand Prince.

He became known as Ivan *Kalita*, "Moneybags," for the wealth he accumulated. Money and power tend to attract more of the same, and he was able to use it to begin the process of expanding Moscow's realms. Some smaller principalities, such as Beloözero and Uglich, he essentially bought; others, such as Rostov and Yaroslavl, he brought into the dynasty's control through marriage. What the Ryurikids took, they would keep. Ivan instituted the practice of inheritance by primogeniture: the entire estate passed to the eldest son rather than being divided into multiple appanages. The family business remained concentrated and thriving.

Ivan's successor Simeon the Proud (r. 1341–53) began to cast his eyes on Novgorod, seizing from it the lucrative town of Torzhok. Ivan II (r. 1353–9) was less successful, not least because at this time Russia was ravaged by the Black Death, killing perhaps a quarter of its population. Ivan was considered weak and passive by the ruthlessly opportunistic standards of the Ryurikids. His son Dmitry (r. 1359–89), by contrast, was daring and imaginative, and would take a gamble that could have proven disastrous, but instead turned out to be the true making of Muscovite dominance over all the Rus'.

Dmitry and Kulikovo

Prince Dmitry was in a very different strategic situation than his predecessors. The Golden Horde was in decline, its vigor diminished, its leaders fighting among themselves, the valuable trade along the Eurasia-spanning Great Silk Road dwindling. Meanwhile, Moscow had reached what seemed to be a peak in its own power, with worrying hints of its own fall to come. The city was now studded by cathedrals and girdled with fortresses. In 1325, Metropolitan Pyotr, the head of the Russian Orthodox Church, had moved his seat to Moscow, symbolically making it, not Kiev or Vladimir, the spiritual capital of all the Russias.

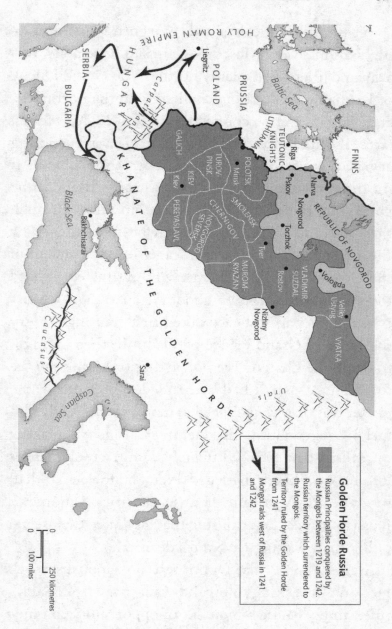

HOLY ROMAN EMPIRE
SERBIA
HUNGARY
POLAND
PRUSSIA
BULGARIA
Liegnitz
Carpathians
Baltic Sea
GALICH
TEUTONIC KNIGHTS
LITHUANIA
Riga
FINNS
POLOTSK
Minsk
Pskov
Narva
KIEV
TUROV-PINSK
Novgorod
REPUBLIC OF NOVGOROD
Kiev
PERESLAVL
SMOLENSK
CHERNIGOV
NOVGOROD-SEVERSK
Torzhok
Black Sea
Bakhchisarai
Tver
VLADIMIR-SUZDAL
MUROM-RYAZAN
Rostov
Vologda
Veliky Ustyug
K H A N A T E   O F   T H E   G O L D E N   H O R D E
Caucasus
Sarai
Caspian Sea
Urals
Nizhny Novgorod
VYATKA

Golden Horde Russia

Russian Principalities conquered by the Mongols between 1219 and 1242.

Russian territory which surrendered to the Mongols.

Territory ruled by the Golden Horde from 1241

Mongol raids west of Russia in 1241 and 1242

0
0
100 miles
250 kilometres

© HELEN STIRLING

Yet not its political capital. Novgorod was increasingly dismissive of Moscow's claims to hegemony. Ryazan and Tver were openly hostile. The Grand Duchy of Lithuania was a rising challenger to the northwest, and Moscow's close connection to the Golden Horde, once a source of wealth and security, was becoming at best less useful, at worst an outright problem. It meant, after all, that the city was more likely to be pulled into the internecine struggles in Sarai. Through the 1360s and 1370s, it had largely been dominated by Emir Mamai of the Jochids, leader of one element of the Golden Horde, a schemer who ultimately proved too cunning for his own good. He tried to play Moscow and Tver against each other, first granting Dmitry the *yarlyk* of Grand Prince of Vladimir, then giving it to Prince Mikhail of Tver when Dmitry didn't pay the full tribute Mamai had demanded. The truth of the matter was that Mamai needed more and more silver to fuel his own political intrigues in Sarai and was making unrealistic demands. Ultimately, Dmitry took matters into his own hands, besieged Tver, and forced Mikhail to grant him Vladimir. It was a significant moment: Russian princes determining that city's fate themselves rather than waiting for Sarai's decision.

At this point, Prince Dmitry was no nationalist rebel. He was not seeking to break Russia away from Sarai's rule, simply to use a moment of opportunity to renegotiate the terms in Moscow's favor. However, Mamai was facing a new and frightening challenger by the name of Tokhtamysh. Mamai was a plotter rather than a general like Tokhtamysh. He needed money to buy armies and allies—which he would have to squeeze out of a reluctant Russia—and he also needed to prove himself a decisive war leader. To this end, in 1380 he issued a demand for even more tribute than usual. Anticipating that Dmitry either wouldn't or couldn't comply, he began mustering a powerful expedition against Moscow, to take his tribute by force and also demonstrate his martial prowess.

Dmitry hadn't wanted a war—his first instinct had been to try and scrape together the money—but when he heard of Mamai's invasion, he decided to make a virtue out of a necessity. If he had to fight, he would turn this war into a rebellion against the Golden Horde and use that at once to try and consolidate Moscow's dominance over the Rus' and rewrite its reputation, turning the quisling city into the vanguard of Russia's independence.

"Grey wolves ran howling from the mouths of the Don and Dnieper," goes the epic *Zadonshchina*, "ready to rush into the Russian land. But these were not grey wolves but vicious Tatars, who wanted to fight their way through all the lands of the Rus'." Mamai had gathered a force of perhaps 50,000 soldiers: Mongol-Tatars, Armenian auxiliaries, Genoese mercenaries from their trading stations on Crimea. Marching to join him were 5,000 Lithuanians under Grand Duke Jogaila and—with some reluctance—a thousand men under Prince Oleg of Ryazan, whose southeastern city was too close to Sarai to risk defiance. Dmitry was able to muster a force more like 30,000 strong, half from Moscow and its subject cities. In part this disproportion reflects the widespread suspicion of Moscow's ambitions by the other princes: Novgorod, Tver and even Dmitry's father-in-law, Prince Dmitry Konstantinovich of Suzdal held back.

The two armies met at Kulikovo, Mamai impatient to secure a victory before winter ended the campaigning season, Dmitry desperate to bring his enemy to battle before he could be reinforced by the Ryazan and Lithuanian contingents. It was a brutal, bloody fight—"men fell like hay under the scythe, and blood flowed like water in streams"—but Russian resolve and cunning finally won the day, Dmitry springing an eleventh-hour ambush that turned Mamai's flank and routed his men. Dmitry lost perhaps a third of his army, but gained wagonloads of loot and, even more importantly, the reputation as the Russian champion who had defeated the irresistible Golden Horde.

Here is the point where myth and reality diverge most sharply. This was an undoubted battlefield triumph, but not a political turning point. Mamai would meet his end in Crimea, killed by the Genoese whose mercenaries he had allowed to die to give him time to flee. Tokhtamysh would consolidate his power and return with a new army, burn Moscow and force Dmitry *Donskoi* to bend the knee. The Russians would continue to be vassals of the Mongol-Tatar khans until they were faced down in the Great Stand on the Ugra River in 1480 by Grand Prince Ivan III (r. 1462–1505), Dmitry's great-grandson. In the intervening century, Moscow would slowly continue what became known as the "Gathering of the Russian Lands," a two-steps-forward-one-back process of consolidating its grip on the various principalities. Nonetheless, princes continued to travel to Sarai to be confirmed in their positions, and dynastic wars and intercity rivalries continued unabated.

The End of Foreign Rule

So much for the reality. The myth popularized at the time, and assiduously developed in the centuries since, was of a conclusive and dramatic victory and one which confirmed Moscow's position not simply as the foremost principality of the Russias but also one whose status was endorsed by God. Dmitry *Donskoi* had, after all, cultivated the church and also made a point of inviting foreign traders with him to Kulikovo, so that they could spread the word of his victory. Although his successors would often face serious challenges, Dmitry certainly averted the impending decline he had feared for Moscow. Now, Kulikovo is a shrine to Russian nationalism, and in 1988 Dmitry was made a saint of the

Russian Orthodox Church. In 2010, Patriarch Kirill said the "battle proved to everyone that Russia was like a powerful coil, capable of springing out and throwing back any opponent and go on to win."

More broadly, though, the era of the so-called "Mongol Yoke" has become central to Russia's imagined picture of itself—and many outsiders', too. The conventional wisdom is that Mongol rule locked Russia away from a Europe that at the time was going through the Renaissance and the early stages of the Reformation. Instead of experiencing the cultural, social, economic and religious changes of those centuries, the poor Russians were lost in what Karl Marx fancifully called "the bloody swamp of Mongol slavery." Meanwhile, the Russians internalized ruthless, "Asiatic" forms of rule, in which absolute power was wielded from the top with absolute brutality, demanding absolute submission from below. Moscow, as the city most closely tied to the Golden Horde, most enthusiastically adopted this political culture, and as it gathered the Russian lands, it also made them an image of itself.

Perhaps. To a degree there is truth here, but only the partial truth of a caricature. First of all, Mongol conquest did not seal Russia behind some "yurt-felt curtain." Traders and emissaries, exiles and missionaries still traveled back and forth. Novgorod maintained its foothold in the Baltic, and Muscovite princes made dynastic marriages with both Constantinople and Lithuania. The difficulties of east–west travel without suitable river routes through the forest and the relative poverty of Russia are probably equal explanations for any isolation. After all, would there have been a Renaissance in Russia had it escaped Mongol invasion? To a large extent this movement, spreading from its epicenters in the Italian and Dutch cities, was the result of improved agrarian yields and thus a burgeoning mercantile class and urban population. The Mongol invasion certainly set back Russian urbanization and the city-based artisanal economy, and the added burden of tribute also had an impact on trade and agrarian expansion. Even so, it is difficult to imagine a Renaissance amidst the deep forests of Russia.

Likewise, some historians have argued that the Russians ended up adopting Mongol styles of rule wholesale. In part, they adduce this from the numerous words relating to governance that Russian has borrowed from them, from *yarlyk* (now used for a custom stamp) to *dengi* (money). However, absolutism is hardly an Asian invention, and the term *tsar*, emperor, which would in due course be adopted by the princes of Moscow, is rooted in the Latin *caesar*, and was actually applied to the Byzantine rulers.

The foundations of authoritarian rule in Russia could as easily be found in Constantinople-facing Kiev as Sarai-beholden Moscow. While there is no question that the Golden Horde had a greater influence on the latter, not least as many of its princes spent years among their Mongol masters, it is a convenient myth for everyone to blame the "Yoke" for a supposed Russian predisposition to despotism. For Russians, the Mongols gave them an alibi. For outside critics of Russia, past and present, this likewise provides a perfect way of "Othering" them, of defining them not as Eastern Europeans but as western Asians, or as some bastardized hybrid at best. "Scratch a Russian," the nineteenth-century French aphorism had it, "and you'll find a Tatar."

The authority of the Golden Horde over the Russians was much more conditional than generally assumed, and often dependent on the support of local princes. Likewise, frequently it was the princes who used Sarai to prosecute their own schemes and advance their own interests. Set aside the devastation of the initial invasion—admittedly a lot to set aside—and the roots of Russian absolutism seem to be found in the objective circumstances of the time and the place. A poor land, in which princes needed to control their cities and their peasants firmly to extract as much tax as they could from them. A land where—away from the Mongols' admirably fast *yam* postal routes—news and orders traveled slowly. The consequent tendency to autonomy demanded especially harsh treatment on the parts of overlords—Mongol or Russian—as a deterrent. To be sure, the Golden Horde, like Constantinople before it, gave them some practices and idioms of power, another layer of writing on the palimpsest. But Russia was still its own country, and Ivan III—Ivan the Great—and his successors were about to have the chance to show just what a country that would be.

Further reading: Janet Martin's *Medieval Russia 980–1584* (Cambridge, 2007) is still the best general textbook on this era, although the first part of Robert O. Crummey's *The Formation of Muscovy, 1304–1613* (Longman, 1987) is worthy of note. Charles Halperin's *Russia and the Golden Horde: The Mongol Impact on Medieval Russian History* (John Wiley, 1985) and Donald Ostrowski's *Muscovy and the Mongols: Cross-cultural Influences on the Steppe Frontier, 1304–1589* (Cambridge, 1998) are the scholarly classics in the field. *The Story of the Mongols, whom we call the Tartars* (Branden, 1996) is a translation of Giovanni DiPlano Carpini's account of his thirteenth-century travels all the way to Karakorum. For those who want to know more about Kulikovo, and what we do and don't know about it, I cover this in *Kulikovo 1380: The battle that made Russia* (Osprey, 2019).

3

"AUTOCRACY, BY GOD'S WILL"

Timeline

Detail of Ilya Repin's Ivan the Terrible and His Son Ivan on Friday
16 November 1581 *(1885)*

It's always hard living down an embarrassing parent—
ask any teenager. For Russia, though, it is awkward to
admit but impossible to ignore that so much that defines it today, from the institutions of the state to its expansion south and east, can be traced back to the reign
of Ivan IV, known as Ivan the Terrible. Admittedly,
a better translation of the Russian *Grozny* would be
Ivan the Dread, or even Ivan the Awesome, however
much this makes him sound like a Californian surfer.
By any standards, he was an extraordinary figure who
built the foundations of the modern Russian state, created a country-within-a-country, unleashed terror on
his own people and even offered Queen Elizabeth I

of England his hand in marriage (it was an offer she could refuse).

Ilya Repin's harrowing picture captures the moment in 1581 when, in a fit of rage, Ivan hit his son on the head, killing him. The brooding master of all the Russias is reduced to dumbstruck horror, his wide-eyed gaze evoking the cycles of paranoia and remorse into which he had become locked in his later years. More than just a personal tragedy, this left the fragile and reclusive Fyodor as his only heir, and thus triggered a series of events that would plunge Russia into a maelstrom of rebellion, invasion, coup and chaos.

From this "Time of Troubles" would emerge the new Romanov dynasty that would rule Russia until 1917, but while many histories of Russia see this as the crucial turning point, in fact the real transition of post-Mongol Russia from a fractious collection of principalities into the Muscovite state took place earlier. Ivan III (r. 1462–1505) began the process, but it was his grandson, Ivan IV (r. 1533–84), who shaped the future of Russia, first as a state-builder, and then as a state-breaker.

The Gathering of the Russian Lands

Everyone stands on the shoulders of others, and Ivan could only be Awesome because his predecessors had been smart, ruthless and focused. Previous Grand

Princes such as Ivan I *Kalita* had started the process of the "Gathering of the Russian Lands" into Moscow's hands, and Dmitry *Donskoi* did much to assert the dynasty's claim to leadership. His son, Ivan III, acquired the soubriquet "the Great," not least for the massive expansion of Moscow's realm. Dynamic and unyielding, he unified the Russian lands through conquest, diplomacy and bribery. Novgorod was finally broken, forced in 1478 to abase itself to its old rival and surrender more than three-quarters of its territories. In 1480, his armies faced down the forces of the Great Horde along the Ugra River, finally ending even the fiction of subordination to the Mongol-Tatars. To the west, he sparred with the Swedes and took cities from the Lithuanians.

At least as important were the changes Ivan III brought to the ideology and institutions of power at home. In 1453, Constantinople had finally fallen to the Ottoman Empire. Moscow's claim to be the "Third Rome," the last bastion of true Orthodox Christianity, became conviction. Ivan, whose second wife was the Byzantine princess Sophia Paleologue, built from this a claim also to be the political heir of the Eastern Roman Empire. Never a man to suffer much from self-doubt and humility, he became increasingly autocratic. The double-headed eagle of Constantinople was appropriated as Muscovy's, and Byzantine court etiquette began to creep into use. Despite overtures from Rome, Ivan slammed the door on any accommodation and the Orthodox Church thrived, with monasteries and cathedrals popping up across the country like mushrooms after rain. With this came a new conservatism. Previously there had been rare examples of women playing serious roles—cosmopolitan Novgorod even had a female mayor, Marfa Boretskaya—but by the sixteenth century, boyars were banishing their sisters, wives and daughters to the seclusion of the *terem*, separate quarters away from the public gaze and the unchaperoned company of men.

Ivan spent lavishly making Moscow, already almost twice as large as Prague and Florence, a fit successor to Tsargrad. He invited Italian architects to expand the Kremlin fortified complex, and to build towers and cathedrals with the tribute now flowing in from his new subjects. Symbolism reflected shifts in real power. Traditionally, the Grand Prince was expected at least to make a show of consulting the boyars, the great lords of the land, but Ivan took to treating them simply as subjects. Although it would be his grandson, Ivan IV, who first formally took the title of *tsar*—emperor—nonetheless this was when the term started to creep into use.

In 1497, the various territories of Muscovy acquired their first standardized legal system with the *Sudebnik*, or Code of Laws. Its subtext was unsubtle and unavoidable: the ever-stronger grip of the Grand Prince on everyone from local officials, who now had less latitude, to peasants, who were now only allowed to move to a new village and master in the two weeks every November on each side of St. George's Day.

When he seized Novgorod's lands, Ivan took the opportunity to create a whole new class of landholding soldiers, the *pomeshchiks*, to whom he assigned small estates off which they could live, in return for military service. Indeed, this became the model for the whole ruling elite, who became bound into the complex hierarchies of the system called *mestnichestvo* (there is no real translation: in effect, it means "place-ism"), which linked status to service to the Grand Prince. A mess of separate and often competing noble families was—in theory—turned into a single service aristocracy. Even princes of subject cities were now not considered royal, their territories no longer theirs to bequeath to their heirs. Autocracy had arrived in Russia, and all those inconvenient traditions of local self-rule and princely independence were consigned to history.

Rising Tsar

Vasily III (r. 1505–33) consolidated his father Ivan III's successes, but his death in 1533 left his son and heir, Ivan, named Grand Prince at the tender and vulnerable age of three. His father's death was only the first in a series of traumas that would shape, or perhaps warp, the man who would become Russia's first tsar. His mother, Yelena Glinskaya, initially ruled as regent in his name, but she died from what was widely assumed to be poison five years later. The regency became a political prize over which the great boyar families of the Shuiskies, Belskies and Glinskies squabbled and feuded, and the neglected child prince by his own account haunted what was meant to be his palace, forced to raid the kitchen for scraps to eat.

In later letters—whose authenticity has, in fairness, been questioned—he railed against the way he and his brother Yuri (who was deaf and thus ineligible for the throne) were treated "like vagrants and children of the poorest." This was a difficult and even dangerous environment, and certainly contributed to a lifelong quest for zones of security, whether physical, political or moral, and an equal lifelong inability ever to feel that sense of ease. On the other hand, it was also a hothouse in which the young prince learned the brutal arts of Muscovite politics fast and well. In 1541, the Khanate of Kazan to the south invaded Muscovy, supported by Ottoman troops. The 11-year-old Grand Prince played no meaningful role in the subsequent Russian victory, but the regents used him as a figurehead and virtual mascot, and thus he also gained some of the credit for success. In a time when omens were seen as very real signs of God's favor (or anger), such symbolic triumphs mattered.

At court, Ivan was beginning to come into his own.

The Shuiskies had become dominant, and they tried to sideline the young prince by surrounding him with rowdy companions keen to distract him with drink and hunting and violent aristocratic pursuits of every kind. He certainly participated, but he did not lose sight of what the arrogant and corrupt Shuiskies were meanwhile doing in his name. In December 1543, while still only thirteen, Ivan ordered the arrest of Prince Andrei Shuisky, and had his kennel-men beat him to death. It was a stark display of the power of the legitimate Grand Prince, and of his determination to rule. The next few years saw Ivan and the boyars in an awkward, often bitter relationship. He needed them to govern his country for him, but mistrusted them, and his erratic swings from denunciation and arrest to conciliation reflected this fundamental tension. He needed some new basis for rule to consolidate his hold on them and—Ivan's perennial quest—bring him security. He found a possible answer in taking his grandfather's reforms one momentous step further.

In 1547, the Grand Prince was crowned Tsar of All the Russias, a symbolic elevation marked with the use in the ceremony of the Cap of Monomakh, a crown that was supposed to have been presented by the Byzantine emperor Constantine IX Monomachus to his grandson, Vladimir Monomakh, founder of Vladimir. This is, of course, myth: Constantine and Vladimir were eleventh-century rulers, and the crown was only made in the thirteenth century. As ever, though, facts take second place when it comes to building narratives of power and authority. Grand princes, for all their wealth and power, had been first among equals, and a robust tradition of egalitarianism had survived the age of the *veche*. Now, the ruler of Russia was to be considered no mere prince or king, but emperor, and with that came a divine mandate, as both defender of the true Orthodox faith and also the Russian people's intercessory with God. Peasant or boyar, soldier or priest, they were all to be subject to a single authority, and one backed by the promise of Heaven and the threat of Hell.

Building a State

Yet Ivan did not rely purely on terror, regalia and a new title, and nor did he necessarily see power as purely an end in itself. This violent and unpredictable man was genuinely pious, and did not take his new role lightly. In the years of regency and boyar infighting, the power of Moscow over the country had waned, and mismanagement had led to local revolts. From the nobility, through the urban traders and crafters, down to the peasants, there seems to have been a general sense of the need for reform, for order, and an end to the kind of cannibalistic competition and exploitation that had become the norm.

He embarked on a series of reforms that were to re-shape Russia, ruthlessly but effectively bringing together the processes started by his predecessors. Under him, the foundations of the Russian state bureaucracy were laid, the laws were codified further, and relations between church and crown defined. In 1549, Ivan addressed a gathering of aristocrats and the church's Sacred Council. He denounced the boyars, but in the name of reconciliation, he reassured them that he would not punish them for past misdeeds. The threat of what he would do if they challenged him in the future, though, hung over them like the headsman's ax. He announced the start of a wide-ranging program of reforms to strengthen and regularize the state. The very next day, he reduced the powers of the *namestniks*, governors who had become virtual local tyrants during his minority. The next year, an updated law code was issued that introduced tougher scrutiny of officials by a royal chancellery. This meant the creation of a central civil service apparatus, virtually from scratch. Here are the roots of the modern Russia state; Ivan's Banditry Office was the forefather of the modern Interior Ministry, for example, while Ivan Viskovaty, founding head of the Ambassadorial Office, is reckoned by today's Ministry of Foreign Affairs to be the country's first foreign minister.

The church did not escape Ivan's reforming zeal. In 1551, religious leaders from across the country met in what became known as the Council of a Hundred Chapters. Reflecting both his wider political agenda and his new status as divinely anointed ruler, Ivan set the agenda with a list of questions as to how they could address abuses by the clergy. The outcome was a document that brought new unity to the Russian Orthodox Church, but bought that by affirming ever more strongly its commitment to the institution of the tsar.

His reforms undoubtedly modernized the country. The boyars were increasingly forced into direct state service and challenged by a new generation of secretaries and *pomeshchiks*. The old practice of allowing officials to sustain themselves by *kormleniye*—"feeding," in other words extorting payments from those under them, whenever and however they felt like it—was banned, and replaced by salaries or, more often, an expansion of the *pomestiye* system of granting land for office. Bit by bit, a fractious and prideful nobility was being forced into becoming a service gentry dependent on the state.

On the basis of Ivan III's earlier efforts, Ivan IV was creating a new kind of monarchy, one that derived its legitimacy from heredity and divine right, yet its power from its ability to represent and balance different estates in society: the boyars, the *pomeshchiks*, the church, the townspeople, the peasants. All were represented in the *Zemsky Sobor*, the Assembly of the Land, which under him, at least, was little more than a rubber-stamp parliament. He was also ambitious and energetic, eager to use the fruits of this campaign to secure the country's borders, and extend them. The irony is that it would be his very success in doing so that would bring whole new threats to Russia's door.

A Creeping Empire

Part of Ivan's vision of a stronger state meant turning a feudal military based on mustering aristocrats' retinues, and thus dependent on their loyalty and efficiency, into a monarchical one. To this end, in 1550 he founded the *streltsy* (literally "shooters"), a force of soldiers beholden to the crown. While the nobility still fought on horseback, the *streltsy* were infantry, armed with harquebuses, early handguns, along with the traditional Russian poleax called a *berdysh*. More to the point, they were neither conscripts nor aristocrats born to their role, but originally volunteers from the towns and the countryside. In due course, service within the *streltsy* would become a hereditary and lifelong right, and their salaries would be supplemented with small plots of land and the right to carry out trade and crafts when not training or on campaign.

On the one hand, this was an expression of Ivan's constant need to try and create security for himself, a military force to guard his Kremlin and police Moscow that was independent of the boyars. However, it also substantially increased Russia's military capabilities, facilitating the expansion of the country's borders in ways both planned and unplanned. The first campaign in which they were used was the successful conquest of the Khanate of Kazan in 1552. Ivan had not forgotten that the Khanate had sought to invade during his minority, and he was determined to deal with this long-running threat once and for all. Russia's traditional skills at woodworking came to the fore with the construction of a fort at Sviyazhsk on the Volga in 1551 in just four weeks from components made in Uglich upstream and floated downriver. Next summer, Ivan launched his offensive, sending an army to besiege Kazan, batter it with fully 150 cannon, and then storm it. The city's chronicle (a possibly questionable source) reported 110,000 killed and more than 60,000 Russian slaves freed.

The south was Ivan's, with the remaining Astrakhan Khanate being annexed in 1556, but he presumably did not appreciate the true significance of this expansion. First of all, it marked the beginning of Russia's transformation from being an essentially homogenous nation, drawn from a single compound ethnicity and sharing the same faith. As it expanded, it came to embrace new peoples, new cultures and new religions, such as the Turkic Muslims of the Khanates. It also brought Russia into direct conflict with the Ottoman Empire, which had its own imperial ambitions in the lands between the Black and Caspian Seas. In 1569, in the first of what would be a centuries-long series of Russo-Turkish wars, the Ottomans launched an abortive attack on Astrakhan. Believing themselves to be facing an imminent Russian threat and under the Ottomans' protection, in 1571 the last remaining Khanate, the Crimean, launched an attack that made it all the way to Moscow's walls. In short, Ivan had looked to end a threat—and made an enemy.

Likewise, to the west, Russia found itself in contention with Sweden, Lithuania, Poland and Denmark over access to the Baltic Sea and its lucrative trade routes. The Livonian War of 1558–83, actually a scrappy, on-off series of campaigns and wars between Russia and one or more of its western rivals, ended in a truce, an uneasy stalemate, and no real victory. Quite the opposite, Russia had lost thousands of men, uncounted treasure and small portions of its territory. More to the point, the rise of the Russian state and the fact that it could contemplate serious military adventures in northern Europe meant that, from being a relatively ignored backwater, it was now considered a serious player and thus a serious threat by some of the heavy-hitting European powers of the age. Ivan had started building an empire—and made himself a threat.

The true expansion of Russia at this time was to the east, into the forests and the steppes loosely claimed by the Siberian Khanate but seen by Moscow as ripe for exploitation. This was essentially subcontracted to ambitious adventurers, most notably the wealthy Stroganov family, which funded a series of expeditions to seize land and set up forts in pursuit of "soft gold"—furs—and the profits in taxing and controlling the trade. Just as with the European conquest of the New World, empire, business, exploitation and taxation advanced together, bureaucrats following the adventurers, as the initial need to enforce tax collection would lead to the need somehow to administer this expanding territory. For now, though, this was the open frontier attracting all manner of renegades and fugitives, mercenaries and explorers, privateers and profiteers. Every year over the next century, Russia would grow by an estimated 35,000 square kilometers (13,500 square miles) on average, roughly equivalent to today's Netherlands or the state of Maryland. Ivan had hoped to find some profit—and found an accidental empire.

Terror and Paranoia

Personalized traditions of rule sat uncomfortably with the new, more European styles of war and governance, though, and Ivan would slip into a murderous state that led to the interregnum of civil strife and invasion known as the "Time of Troubles." For whatever reasons—his traumatic early life, the pain caused by the bone diseases that were slowly crippling him, clinical paranoia—his quest for security would take increasingly erratic and destructive forms.

In 1560, his first wife, Anastasia Romanovna, died, removing a moderating influence on him; Jerome Horsey, a trader with the English Muscovy Company, observed that "she ruled him with admirable affability and wisdom." Ivan appears to have suspected she was poisoned, like his mother. Meanwhile, the Livonian War was going badly, resistance to reform was continuing and, in 1564, one of his closest advisers, Prince Andrei Kurbsky, defected to Lithuania. The tsar's childhood-deep suspicion of the nobility metastasized.

So Ivan decamped to the fortified township of Aleksandrova Sloboda and in effect announced his abdication, blaming the boyars for their "treasonous deeds" and corruption, and the church for covering up their sins. This was a daring challenge to the elites, who had no alternative ruler to replace him and faced the anger of the Muscovites. Fearing invasion from abroad, lynching at home and the country's descent into civil war, they capitulated and begged Ivan to return. He agreed—but his price was the unfettered right to punish whomever he considered a "traitor." In effect, he demanded and was granted absolute power.

Ivan was not going to rely on the boyars' promises, though. He decreed the creation of a state-within-a-state, known as the *Oprichnina* ("Exception"). Largely drawn from the territories of the former Novgorod Republic in the north, he took it as his personal realm. The rest of Russia, known as the *Zemshchina* ("the Land"), he left in the stewardship of the Boyar Council. Within his new territory, Ivan raised a force called the *Oprichniks* to be his personal guards and enforcers. The *Oprichniks* would be unleashed into the *Zemshchina* to purge noble clans who aroused Ivan's anger and even conduct a month-long orgy of massacre and rape in Novgorod in 1570.

The *Oprichniks* wore black monastic robes (Kurbsky called them "children of darkness") and bore a dog's head and a broom, symbolizing that they were the tsar's hounds and would sweep away his foes. They were as ruthless and exploitative as any private army could be. Increasingly, the tsar himself could hardly control them, and they raided and plundered with enthusiasm and impunity. Peasants fled the lands they controlled or attacked, leading to food shortages and a trade crisis, and the tsar himself began to feel prisoner of the very force meant finally to protect him.

When, in 1572, the Crimean Khanate's forces almost took Moscow, it brought home the risks inherent in dividing the country in two. Given that he was already concerned that the *Oprichniks* were out of control, Ivan abolished the *Oprichnina* as suddenly as he had created it, and returned to rule from Moscow. Hopes of a return to the old balance between tsar and boyars were soon dashed, though. He continued to rely on cronies rather than aristocrats, to see plots and treason on every side, and to suppress them as bloodily as ever. His victims were hanged and beheaded, cut to pieces or sewn into bearskins and ripped apart by dogs, but still Ivan saw yet more plotters on every side.

Ivan's reign closed in crisis and confusion. He so mistrusted his commanders that, in the last years of the Livonian War, he attached his own representatives to shadow each of them, like distant harbingers of the Soviets' political commissars. Suspicion and desperation divided the aristocracy. The economy was shattered by war, taxation, banditry and depopulation. Fertile land was going unplowed because of a lack of farmers, who had starved or fled south and east, out of Moscow's reach. So great was the need for peasants that landlords would kidnap them from each other's lands. Meanwhile, the death of Prince Ivan at his father's hand in 1581 had left as heir young, churchy Prince Fyodor, a man manifestly without what it would take to bring this country together—after Ivan the Terrible, Russia would get Fyodor the Bellringer.

This state of affairs could not endure. Venetian diplomat Ambrogio Contarini had been amazed at the markets that sprang up on the frozen Moskva River during the long, harsh winters. He was especially struck by the spectacle of butchered livestock stacked by stalls, deep frozen for weeks or even months. "It is curious," he wrote, "to see so many skinned cows standing upright on their feet." By 1584, the system Ivans III and IV had put time and blood into creating seemed much like one of those cows: dead, skinned, still standing because it was frozen, but ready meat for the butcher's ax.

The Time of Troubles and the Rise of the Romanovs

The Time of Troubles

Internal conflict
— Boundary of Russia 1598
▪▪▶ Attack by "The False Dmitry" 1604–1605
░ Area of uprising of Volga (non Slav) peoples 1606–1608
⋯ Area of Bolotnikov uprising
➤ Bolotnikov's campaign 1606–1607

External attacks
→ Polish siege of Smolensk 1609–1611
••••▶ Attacks by Swedes on Novgorod 1610
→ Polish campaign against Moscow 1610–1613
→ Russian counter-attack 1612
—▪— Area occupied by Poles 1612–1613
⋯⋯ Area occupied by Swedes to 1613
▨ Ceded to Sweden by Peace of Stolbovo
➤ Wladislaw's campaign against Moscow 1618
▨ Ceded to Poland by Peace of Deulino
— Boundary of Russia 1618

Cossack leader Ivan Bolotnikov raises a peasant revolt against Moscow

© HELEN STIRLING

In 1584 Ivan died, laid low by a stroke in the middle of a game of chess. Pious, naive Fyodor (r. 1584–98) was duly crowned, but real power was in the hands of the Romanov and Godunov families, and especially Fyodor's brother-in-law, Boris Godunov. Once again, the court became a battlefield between clans. Godunov purged the rival Belsky family in 1584, then the Shuiskies and the Nagoi in 1587. Meanwhile, Tsar Fyodor occupied himself with visiting churches across the country and organizing bellringing.

The country lurched from crisis to crisis. In 1590, Godunov started a war with Sweden, hoping for some quick gains. Five years later, the Treaty of Tyavzino left Russia with little to show for the cost. Peasants continued to try to flee the land, and agrarian collapse led to a vicious circle of banditry and scarcity. When Fyodor died in 1598, childless, the Ryurikid dynasty was over. Godunov seized the opportunity and his ally Patriarch Job nominated him to the *Zemsky Sobor* as successor. Whether out of fear or genuine conviction, they approved him unanimously.

Godunov (r. 1598–1605) was crowned as new tsar. A former *Oprichnik*, he was smart and ambitious, ruthless and competent. But he had been elevated not by God, but by men, and all his qualities seemed to count for little compared with that handicap. It did not help that his reign was marked by famine, interpreted as a sign of divine disfavor, and an associated peasant rising. In

1604, a pretender claiming to be Dmitry, Fyodor's half brother—who had actually died in 1591—led a Polish-backed attempt to seize the throne, and Russians, excited by the thought that maybe the Ryurikid dynasty had survived, flocked to his ranks. "Fake news" was already destabilizing governments in the sixteenth century.

When Godunov died in 1605, his 16-year-old son Fyodor II reigned as tsar for just two months before being murdered. After all, without divine right, one candidate's claims were just as good as another's. The next eight years are known as the Time of Troubles. The False Dmitry was acclaimed as tsar and himself soon killed. There were coups and intrigues, rebellions and risings, another False Dmitry and a Polish invasion.

Ultimately, this was the culmination of three, long-term processes. It was a dynastic crisis: having established a notion of a sacral ruler legitimated by heavenly mandate, the system could not accommodate a dynastic break, especially as a willful and ambitious aristocracy were emboldened by this in their struggle against an emerging centralized autocracy. It would take the Time of Troubles to break them. It was also a socio-economic crisis, as hereditary boyars squared off with the service gentry, and the flight of peasants from the land undermined them both. It would take the Time of Troubles to force the regime to address these challenges squarely, in effect making both servants of the state. Finally, it was a geopolitical crisis. As Russia rose, it found itself facing new and more formidable threats: Crimean Tatars and Ottomans to the south, and above all the Poles and the Swedes to the west. It would take the Time of Troubles to turn Russia into the kind of modern tax-raising and army-building machine that was emerging in Europe.

Scratch a Russian, Find a…Byzantine?

There is much scholarly debate on whether or not the political culture and institutions of Russia at this time were Mongol-Tatar ones, lightly gilded with a veneer of Byzantine pomp and pageantry. How far does that really matter, though? Russian Orthodox saints were often pagan gods given a halo and a new backstory. Track the family trees of Russia's great families and they were typically a mix of Slav, Varangian and Tatar. The traditional *veche*, the city assembly, was derived from ancient Slavic practice, but merged with the *thing* gatherings familiar to the Vikings. The point is not where different ideas and practices came from, but how they were conceived, what meaning people took from them, and how they developed a life of their own, shaping the country and its people in the future.

Before, Russia had been a canvas on which successive cultures had written their ideals and intentions. Some of these cultural inscriptions lasted, were embellished and emphasized by later generations; others were soon painted over. But the point was that until the rise of Muscovy, the Russians' role in this process had largely been reactive, even passive. Now, though, they were actively seeking to define themselves, and for that they looked abroad.

Tsar. The new title emulated that of the Roman emperors, and it was the double-headed eagle of Eastern Rome that the ruler took as his symbol. Like the notionally divine emperors of Rome and Constantinople, the tsar was a sacral sovereign, subject only to the God who also gave him his mandate. As Ivan put it in one of his intemperate letters to Prince Kurbsky, the defector, Russia now had "autocracy, by God's will," and he was "the Orthodox, truly Christian autocrat."

Austrian diplomat Baron Sigismund von Herberstein wrote of Ivan the Terrible that the boyars, "being either moved by the grandeur of his achievements or stricken with fear, became subject to him." The latter was more likely the case, but fear of an individual can be a flimsy basis for lasting power. The two Ivans created the ideological, institutional, even aesthetic basis for divine-right autocracy in Russia, but it would take the comprehensive complex of crises of the Time of Troubles to make Russians, from peasants to boyars, gratefully accept such a ruler, as the alternative to chaos, hunger and invasion.

Eventually, in 1613, the *Zemsky Sobor* offered the crown to 16-year-old Mikhail Romanov. They wanted a tsar, needed a tsar, and ended up having to create one.

His main qualifications, after all, appeared to be that he was unobjectionable, from a family able to trace its history back to the days of the Kievan Rus', and the son of the formidable Patriarch Filaret. But the truth was that an exhausted Russia demanded a stable future and Mikhail was able to provide that, reigning until 1645 and ushering in the dynasty that would rule Russia until 1917.

Further reading: Andrei Pavlov and Maureen Perrie's *Ivan the Terrible* (Pearson, 2003) and Isabel de Madariaga's book of the same name (Yale University Press, 2006) are the best biographies of this complex man. Robert O. Crummey's *The Formation of Muscovy 1304–1613* (Longman, 1987) is a dense read, but rich in detail. For a change of pace and tone, Vladimir Sorokin's *Day of the Oprichnik* (Farrar Straus Giroux, 2011) is a short, scatological science-fiction book that imagines the *Oprichniks* once again ascendant in a near-future tsarist Russia.

4

"MONEY IS THE ARTERY OF WAR"

Timeline

Zurab Tsereteli's Peter the Great (1997)

Towering 98 meters (322 feet) above the Moskva River, between the hipster bars of the former Red October chocolate factory, the posh houses on Prechistenskaya Embankment and the statue gardens of the Muzeon Arts Park, stands Tsar Peter the Great (r. 1682–1725), immortalized in a thousand tons of steel, bronze and copper. It is a gloriously ugly monument of the man atop a galleon, erected in 1997, during the time when Yuri Luzhkov was mayor of Moscow. In between approving the demolition of historic buildings so that they could be replaced with tacky shopping malls,

Luzhkov commissioned it from Zurab Tsereteli, his favorite sculptor and architect. Most Muscovites loathe it. For a start, it was probably not intended to be Peter. Although Tsereteli now denies it, the prevailing view is that this was actually a design intended to mark the 500th anniversary of Christopher Columbus's first voyage to the Americas in 1492. When no suitably gullible and tasteless American sponsor could be found, Tsereteli simply swapped out the heads, and pitched it to Luzhkov as a statue of Peter, to mark the 300th anniversary of his founding the Russian navy. The rest is history.

Except what kind of history? On one level, it is deeply ironic that Moscow has ended up with such a monument to a tsar who so disliked the city that he built himself a new capital to the north—St. Petersburg. (When, after Luzhkov was forced from office, Moscow offered it the statue, the St. Petersburg city council replied that they didn't want to "disfigure a great city." They meant their own.) Secondly, that the Russian "Peter" is actually the Italian Cristoforo Colombo under a thin disguise is also quite a powerful metaphor for many of the reforms he imposed on Russia. He was a modernizer, a Westernizer even, but instead of seriously tackling the underlying reasons why Russia was different, many of his measures simply addressed the superficialities. Russian aristocrats, for example, were forced to cut their bushy beards or else pay a special tax. But affecting European-style shaved chins doesn't necessarily bring European-style thinking.

Thirdly, though, the very focus on Peter the Great as one of the defining figures of Romanov Russia is understandable. Much like the statue, he towered over those around him both figuratively and literally. He was a veritable giant, over 2 meters tall (6 foot 8) in an age when the average man was 1.7 meters (5 foot 6). He had a prodigious enthusiasm, forever seeking to learn new skills, from dentistry (his unfortunate courtiers had to let him practice on them) to clock-making. He had a genuine curiosity for the outside world, even traveling across Europe, a first such venture for a Russian ruler. Nonetheless, in many ways Peter was at most a culmination of a process. Many of his reforms were rooted in the practices of his Romanov predecessors, and his policies were often dictated not by his own will but by the circumstances in which he found himself.

Finally, just like the twist in fate, taste and patronage that saw the largest monument to Peter in the city he despised, all this was steeped in paradox. He was a Russian nationalist who ordered his own aristocrats to look more European, who adopted ideas and technologies from all across the West, and yet who in effect codified Asiatic despotism, making service to the state the sole basis of status. The more he tried to cherry-pick the most appealing or useful aspects of Europe for Russia, the more he had to find ways to justify this in terms of a Russian divine mission and special place in the world. This was the greatest fiction of all, that modern European culture at the top of the system could coexist with Eurasian feudalism below, best epitomized by the building of his new capital, St. Petersburg. An airy, modern city designed by French and Italian architects—and built by half a million serfs dragooned from across the country, tens of thousands of whom would die.

Enter the Romanovs

From the Time of Troubles emerged not only the new Romanov dynasty, but also a new, cohering narrative: that Russia would be prey for its many foes if it did not have a single, powerful ruler around whom all the classes and peoples of the nation could—and must—unite. This became the basis for the Russian Empire, and with it a growing national self-image as both beleaguered fortress amidst a sea of enemies and also guardian of everything that was good and proper, from the true faith to, later, legitimate regimes facing the chaos of anarchy and rebellions. With this, though, came the inevitable tension: how to secure the borders, assert Russian interests and maintain order at home, without adopting Western technologies? And could those technologies be adopted without associated social and even political changes? The answer was ultimately that they could not, but for centuries the tsars would certainly try. This would thus be a time of growth and strength mixed with peril and paradox, and the seventeenth century would be one of wars abroad and risings at home, but also of imperial expansion and growing national self-confidence.

Mikhail (r. 1613–45), first of the Romanov tsars, may have been chosen for his blandness, but his reign proved unexpectedly productive (in part, admittedly, because of the role played by his domineering father, Filaret). His actual coronation was delayed by weeks because Moscow, battered by successive wars and rebellions, stripped and starved, was in too bad a state to accommodate it. By the time of his death in 1645, though, he had secured peace with Sweden and Poland, reorganized some of the army along Western lines (something that would lead to rebellion by the traditional *streltsy* under Peter) and overseen the expansion of Russian influence into Siberia by a motley and often murderous array of Cossack mercenaries, fur-trading merchant-adventurers and foresighted aristocrats. In 1639, a band of Cossacks even reached the Pacific Coast, while behind them would come stockades, tax collectors, missionaries and smallpox, decimating the thinly scattered indigenous population of Siberia more viciously than any gun or blade.

The challenge would always be how to balance

the urge to expand and compete with maintaining stability at home. After Mikhail, Alexei (r. 1645–76) acquired the soubriquet "the Most Quiet" for his subdued manner, but faced a tumultuous time of wars with traditional enemies Poland and Sweden and new challenger Persia, as well as a Cossack rebellion that saw towns burn along the Volga and a short-lived Cossack republic, followed by the Pereyaslavl Accords that saw the largest of their communities—and with it much of what is now Ukraine—brought under the rule of the tsar. A schism ripped at what had seemed one of the most stolid and solid of institutions, the Russian Orthodox Church. Alexei was clearly bedeviled by the usual Russian dilemma. On the one hand, he resented the growing influence of foreigners, and their new and different ideas. In 1652, for example, he set up a separate part of Moscow, the "German Quarter" (the Russian word for German, *Nemets*, being used for all foreigners), as a ghetto into which to confine their embassies, their mansions and their churches. In 1675, he banned his court from affecting Western clothes or styles, even in private. Just as wealthy Russians were excited by the exotic and the foreign, though, the Russian state needed these outsiders, their money, their technologies and their military experience. Notably, Alexei may have despised Western ways, but he made Patrick Gordon, a

Roman Catholic Scottish mercenary, a tutor for his son, Peter, one of many foreigners who played a crucial role in shaping the passions and interests of the young *tsarevich* (prince).

Belief and Believers, Old and New

This tension was especially evident in the church. If for the secular authorities the lesson of the Time of Troubles was that weakness at home meant vulnerability abroad, within religious circles there was a growing body of opinion that viewed that era as evidence of God's dissatisfaction with the Russian people and the impurity of their liturgy. In 1652 one of their number, the eloquent and forceful Nikon, became Patriarch of Moscow and All the Rus'. By all accounts he had been reluctant to take on this position, but once he had, he plunged straight into reforms intended to, as he saw it, purify a church that had deviated too far from its Greek Byzantine origins.

Contemporary Greek rites and liturgies replaced those in place (if you valued your life and freedom it was best not to mention the irony that, by then, Russia's were actually closer to those of old Byzantium). Newer styles of icons were banned and Nikon's followers broke into churches and houses across Moscow to seize and burn them. Those who painted them—and the evolution of the artistic styles of these paintings of saints and religious scenes had been one of the glories of early modern Russian culture—had their eyes gouged out and were then paraded through the city. Churches believed to have deviated too far from Byzantine standards were demolished. Even how the name Jesus was to be written and the exact way in which the sign of the cross was made were revised. Former allies who were horrified by the direction and severity of these changes were excommunicated. Piety, Nikonian-style, was imposed by violence, fear and synod courts.

Tsar Alexei had long been entranced by Nikon—he had gone down on bended knee in 1652 to beg him to take the position in Moscow—and at first he was virtually the tsar's right-hand man and stand-in. In the early years of the off-and-on First Northern War against Poland and Sweden, which began in 1654, when Alexei was away at the front, Nikon was effectively regent in Moscow. Over time, though, this relationship would become increasingly tense. Whatever Nikon may have said about bowing to the secular authority of the crown, conversely he thought the crown ought to genuflect to the church on matters spiritual. This included going against the terms of the *Sobornoye ulozheniye*, the new law code issued in 1649, which undercut the authority of the church and reduced its privileges.

Facing resistance from boyars and clergy alike, and a new distance from the tsar, Nikon tried to take a leaf out of Ivan the Terrible's book, symbolically tossing aside his patriarch's robes and abandoning Moscow for a monastery. He waited for his critics to come to their senses and beg for him to return. He waited in vain. For eight years, Nikon and the church remained at an impasse, until the Great Moscow Synod of 1666 in which a conclave of the most powerful clergy and respected theologians finally gathered—in some cases, it is said, induced to do so by generous payments in rubles and furs—to try and settle the crisis. The synod squared this circle by damning Nikon, stripping him of his authority and sending him to a distant monastery under guard, while accepting his reforms. In the great schism known as the *Raskol*, the sundering, those traditionalists who resisted these changes, the so-called Old Believers, were declared apostate and would be persecuted for most of the next three centuries: only in 1971 would the Moscow Patriarchate finally lift its proscriptions on them.

Debates over the precise hand gestures used to cross oneself may seem trivial, and hardly the reason for generations of rancor and sectarianism, murder and exile. However, the religious debates of the Nikonian era reflected a wider fear that, bit by bit, Russia was losing its traditions, its unique place in the world, its soul. The irony is that the "reformers" were seeking to return Russian spiritual life to something it had never been: they mistook contemporary Greek rites for true

Byzantine ones, and were trying to "recreate" a perfect separation of church and tsar that would have been recognized by neither an emperor in Constantinople nor a prince in Kiev. Once again, appeals to history were actually invoking artful (if probably unknowing) reinventions of Russia's past.

Two Tsars for the Price of One

Meanwhile, the secular state continued to edge toward modernization. Alexei's immediate successor, Fyodor III (r. 1676–82), established the Slavic-Greek-Latin Academy, Russia's first institution of higher learning—almost 600 years after the founding of the universities of Bologna and Oxford. Perhaps most striking was his abolition in 1682 of the *mestnichestvo* system, which meant that aristocrats' positions were defined by birth and status. Instead, he encouraged a more meritocratic system in which jobs went to those best suited to them (or by royal appointment: even a reforming tsar was going to have his favorites). The old pedigree books, painstakingly encyclopedic genealogies that were used to determine every aristocrat's precise position in the hierarchy—seating someone a little too low down the table than their status demanded could even trigger a duel—were symbolically and ceremonially burned.

Fyodor died that same year. He left no heir so, in theory, the next in line was his younger brother, Ivan, last surviving son of Alexei's first marriage. The 15-year-

old Ivan was a chronic invalid, though, and according to many equally disabled intellectually. The boyars
feared what could happen if Russia had a weak tsar and
instead looked to his younger half brother by Alexei's
second wife, nine-year-old Peter. However, they had
not considered the traditional rivalries between the
Miloslavsky and Naryshkin families, from which had
come Alexei's first and second wives, respectively, and
the ruthless passions of Ivan's older sister Sophia Alex-
eyevna. Russia might not have been ready for an empress, but Sophia was ready for the next best thing.

She and the other Miloslavskies stirred up a rebellion
among the *streltsy*, spreading rumors that Fyodor had
been poisoned and Ivan strangled. Already angry at the
erosion of their privileges and the rise of new, Western-
style regiments, the conservative *streltsy* were easy to
turn. As the Moscow mob seized the opportunity to
riot and loot, the Boyar *Duma* (assembly) scrabbled for
a compromise. As ever, pragmatism won out, and was
then hurriedly robed in the mantle of invented tradition. Ivan (r. 1682–96) and Peter (r. 1682–1725) were
crowned as Russia's *dvoyetsarstvenniki*, double tsars, with
Sophia as regent. A special double throne was built for
the two youngsters, a duplicate of the ceremonial Cap
of Monomakh hurriedly made so each had one for the
coronation, and Byzantine ritual plundered and heavily
edited to justify such an unusual move.

For just over six years, Sophia, supported by her

ally and maybe-lover Prince Vasily Golitsyn, ran the country. Ivan dutifully spent his days in prayer, pilgrimage and court pomp, and Peter spent his time at the royal estate of Preobrazhenskoye, not least setting up his so-called "play army." A band of retainers and fellow teenagers, over time it became a genuine force a hundred, then three hundred strong. Sophia's regency saw the signing of the unrealistically optimistically named Eternal Peace Treaty with Poland (1686), which ratified Russia's possession of the ancient capital of Kiev, and at the other end of the sprawling empire, the Treaty of Nerchinsk with China (1689). It also saw disastrous campaigns in 1687 and 1689 against the Crimean Khanate in which Russia was defeated not so much by enemy force of arms as the logistical challenges of mounting military expeditions at the borders of what had become such a large state.

Meanwhile, Sophia either could not or would not declare herself *tsarina*, empress. Instead, she had to watch as Ivan sickened and Peter became increasingly willful. In 1689, the 17-year-old Peter had decided that enough was enough. He demanded that Sophia step aside. Though she again sought to raise the *streltsy* against him, she faced the majority of the boyars, most of the *streltsy* and Peter's "play army": by then, two fully fledged companies, with their own cavalry and artillery. Perhaps equally importantly, even Ivan was willing to go along with Peter.

So Sophia was forced into the Novodevichy monastery, something of a traditional genteel jail for unwanted aristocratic womenfolk, from Ivan the Terrible's daughter-in-law to Boris Godunov's sister. Perhaps the ever-planning Sophia had predicted even this fate, as she had made a point of having the ironically named "New Maidens' Monastery" renovated during her regency. And while Peter was technically still a ward of his mother until he was 22, and co-monarch with Ivan until the latter's death in 1696, in practice he was now the tsar. Power was his; but what did he want to do with it?

Building the Petrine State

Much is known about Peter; much less is truly understood. He was charismatic and energetic, but suffered from seizures and facial tics. His motto was "I am a student and I seek teachers" and he was certainly willing to be taught—he did not lead his "play army," but enrolled as a mere bombardier so as to learn warfare from the ground up—but his interests were in matters not intellectual but practical. He was proud of his country, but more desperate for the respect of foreigners than of his own people. He had struggled for power, but once he had it, he seemed uninterested in many aspects of rule, instead indulging himself in the duties that pleased him, neglecting those that did not. Peter had played as soldier when still a boy, had been terrified by the brutal rising of the *streltsy* in 1682 (his uncle Ivan Naryshkin and Artamon Matveyev, the statesman who had proposed his coronation, were both hacked to death before his eyes), risen to power thanks to his personal army and seen Sophia's legitimacy critically undermined when her favorite, Golitsyn, failed in not one but two Crimean expeditions. Military power was, to Peter, vital for his own security, central to Russia's and also, frankly, fun.

He may have neglected the ceremonial that had occupied so many tsars' lives and paid only lip service to any spiritual role. However, he had an intensely practical perspective on statecraft and realized that military strength was based not just on the valor of a country's soldiers but the quality of the technology, logistics and governance behind them. Whether or not he could be considered a modernizer in every sense is open to debate, but he felt an urgent, passionate drive to make of Russia a great power, a respected power, and that meant wars, wars that he wanted to win. At the time, Russia was not considered a serious military player. Reflecting its place in the Western worldview as something no longer Asiatic, but not yet European, the Austrian envoy Johannes Korb acidly noted that "none but the Tatars fear the armies of the tsar."

Peter wanted to change that, but that would cost: "Money," he noted, "is the artery of war." Even by the standards of early modern states, in days when social security meant charity at best, starvation at worst, the

Russian state became in many ways simply a support mechanism for military forces. It has been estimated that by 1705, the share of the central budget they consumed was anything from 65 to 95 percent. This demanded a working bureaucracy, an efficient tax system, and a more disciplined and professional state machine. Peter set out to build them through a sweeping series of reforms.

Serfdom, so long a challenge for Russia's rulers, was made more inflexible, for the state depended on the peasants working, building and fighting. Hundreds of thousands of peasants—out of a total population of perhaps 7.5 million—would be conscripted for Peter's wars and his construction projects, so they could not be allowed to flee south or east. New fines were introduced for concealing runaways, while from 1724, peasants were not even allowed to travel from their own district without a passport. Meanwhile, new taxes rained down on them, on everything from beehives to cucumbers.

Not that the nobility were exempt from being forced to bend to Peter's demanding program. The abolition of *mestnichestvo* had begun to change the way status and position in Russia were determined, but in 1722, Peter introduced the Table of Ranks, which represented a fundamental revolution in the foundations of Russian aristocracy. Henceforth, all noblemen who wanted to rise within the system had to move up the 14 ranks by service, promotion and ability. Of course, favoritism,

wealth and birth would still have an impact in practice, but the theory was now that the nobility held their
place, power and privileges only insofar as they served
the state. Equally important was that state functionaries promoted to a certain rank acquired noble status;
become a collegiate assessor in the civil service, for example, or a prime major in the Life Guards, both 8th
Rank positions, and you became a hereditary aristocrat. Under *mestnichestvo*, status determined your job.
Now, your job determined your status—and in their
own comfortable way, the nobility were effectively
made serfs of the state.

This even applied to the church. For example, an
igumen—an abbot—was a 5th Rank position, equivalent
to a state councilor or a brigadier. Nikon had sought to
clarify the hazy overlap between church and state by
making the former independent of the latter. Peter's so-
lution was the opposite, in effect to make the church another department of government. This was not just about
power, but money. The church held vast lands and generous tax exemptions, and Peter's wars, his nascent navy,
his reorganized army, all of these greedily demanded
funds. Church properties came under the control of the
state, which eagerly squeezed the church for cash. Much
of the revenue earned from this most conservative and
even xenophobic of bodies would be spent not just on
reforms, but on reforms inspired by foreigners.

Peter on Tour

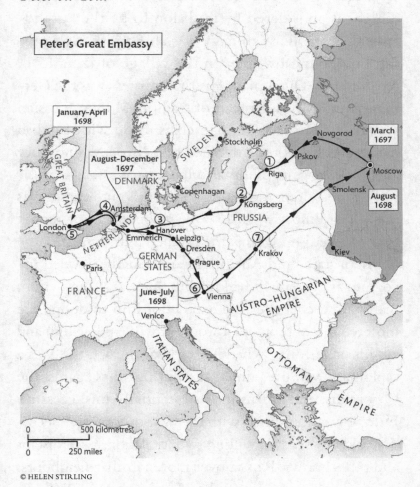

Peter's Great Embassy

January–April
1698

August–December
1697

March
1697

August
1698

June–July
1698

GREAT BRITAIN

SWEDEN

DENMARK

Stockholm

Copenhagen

Novgorod

Pskov

Riga

①

②

Köngsberg

PRUSSIA

Moscow

Smolensk

Kiev

London

⑤

④ Amsterdam

③

Hanover

Emmerich

NETHERLANDS

Leipzig

Dresden

Prague

GERMAN
STATES

Paris

FRANCE

⑦

Krakov

⑥ Vienna

AUSTRO-HUNGARIAN
EMPIRE

OTTOMAN

EMPIRE

Venice

ITALIAN STATES

0   500 kilometres
0   250 miles

© HELEN STIRLING

From the first, Peter had been fascinated by outsiders.
Along with the Scotsman Patrick Gordon, the Swiss
mercenary Franz Lefort was one of the abiding influences of his childhood. (To this day, a neighborhood
in southwestern Moscow close to where the German

Quarter used to be is called Lefortovo, in his honor.)
The boat considered by tradition to be the "grandfather" of the Russian navy was an English-style sailing boat found by Peter in the village of Izmailovo,
repaired by a Dutchman. His first mistress was the German Anna Mons. Instead of traditional long Russian
caftans, he favored German coats made with English
cloth, and in 1700 he decreed that Moscow's nobility
and civil servants ought also to dress Western-style.

The most dramatic expression of this enthusiasm was
the "Great Embassy." In 1697, he set out on an 18-month
tour of Swedish Livonia, the Netherlands, England, the
German states and Austria. He notionally traveled in-
cognito as "Peter Mikhailov," but this was a thin fiction,
really just a characteristic excuse for him to dodge the
most tedious of protocol, carouse at will (and he did) and
get his hands dirty when he wanted. In part, this was a
diplomatic venture, aimed at securing allies against the
Ottomans, and at this it was essentially unsuccessful.
Europe was consumed with the War of Spanish Succession and few seemed willing to put relations with distant
and little-known Russia over more familiar neighbors.

Yet for the curious (and self-indulgent) tsar, it was
also an unparalleled opportunity to explore the West,
its ways, technologies, vices and virtues. In Holland,
he studied shipbuilding and hired the naval architects
who would help build his new navy (tellingly, many
Russian words relating to shipping and the sea have

their roots in Dutch). In England, he likewise sought to learn the tradecraft of naval power and modern monarchy, affirming that it was "happier to be an admiral in England than a tsar in Russia." Nonetheless, having seen parliament sitting, he concluded that "English freedom is not appropriate" for Russia.

Meanwhile, Russia was being governed by Peter's trusted right-hand man, his "Prince-Caesar" Fyodor Romodanovsky, as if the tsar were still in the country. In 1698, though, news came of a fresh rebellion of the *streltsy*. Peter hastened home, although Romodanovsky had suppressed it readily before his return. Nonetheless, Peter was decisive in his response: the *streltsy* were finally disbanded (they were an antiquated obstacle to his creation of a Western-style army) and over a thousand of them were whipped with the savage Russian leather knout, or broken on the rack, or toasted on iron griddles before being hanged or beheaded, their bodies staked out as a public warning: this was a tsar who would brook no disobedience.

The same authoritarianism was evident in his campaign to apply the lessons he had learned in the West. In 1703, after his armies had captured the Swedish fortress of Nyenskans at the mouth of the Neva River, Peter saw the opportunity to build himself both a seaport for the navy he was building and a capital to allow himself to get away from Moscow and demonstrate that Russia could raise a European-style city. His new capital, St.

Petersburg, did indeed turn out to be such a city, built with an eye to Dutch and English planning and laid out by Italian, German and French architects. Yet his "window onto Europe" was built with very Russian methods: tens of thousands of serfs, convicts and prisoners of war were worked to death to give Peter what he wanted. Time itself would bend to the autocrat's will: in 1699 Peter decreed that Russia abandon the Byzantine calendar, which counted the years from the purported creation of the world, to the Western one based on Christ's birth. Thus, the 7207th year suddenly became 1700.

Peter at War

Ultimately, though, beards and architecture, church politics and administrative reform were all about war, the modern army and navy that Peter was so desperate to build—and to use. Of the 28 years in which he was sole tsar, 24 were spent at war between the 1700–22 Great Northern War and the 1722–3 Persian Campaign.

In 1698, a major restructuring of the army saw it modernized and expanded. Every year, one peasant was conscripted from every 20 peasant households, serving for life, so when a young draftee left to join the ranks, he was sent off with a funeral service. There were efforts to standardize equipment and professionalize the officer corps (again, often by hiring foreigners), and a renewed push to produce modern cannon, a passion that dates back to Ivan the Terrible's campaign against

Kazan and was reflected in the later Soviet dependence on its so-called "red god of war."

Peter, obsessed with sailing and shipcraft, realized his dream of, for the first time, giving Russia a navy. By his death, he had built a fleet of 32 ships of the line and more than 100 other vessels. Then again, arguably Russia had not needed a navy before. Only as it began engaging more extensively in maritime trade and contesting northern European hegemony with seafaring powers such as Sweden did this become a priority.

Contest it he did, though, and the Great Northern War was more a string of wars in which combatants came and went, while Russia and Sweden remained the anchors of their respective coalitions. It often had all the elegance of a pub brawl. First Russia was allied with Poland, Denmark–Norway and Saxony, though the latter two had to take a breather for a while until the Swedes lost a major engagement at Poltava. For a while, the Ottomans took the chance to help Sweden give their old enemy Russia a bloody nose, while Hanover and Prussia joined Russia. The British, ever the opportunists looking to prevent any one European nation becoming too powerful, at different times actually supported both sides.

Russia won—or at least Sweden ultimately lost—and established itself as a major power. The Swedish Empire was broken as a military superpower, especially given its relatively small population (of the 40,000 soldiers Charles XII, the "Lion of the North," marched into Russia, only 543 came home). The double-headed eagle had humbled the lion, and shown that it could not just win a battle but build the kind of military and logistical infrastructure that could sustain a long, hard war. Likewise, the 1722–3 Russo-Persian War showed that an army that had not been able to keep a force in the field against Crimea, could now strike deep into the Caucasus and Caspian regions. With the Persian Empire in decline, Peter needed to prevent the Ottomans taking advantage of the situation and extending their control along Russia's southern flank.

So this was Peter's true legacy. He was a war-fighter rather than a state-builder, but he came to learn that to be one, he would need to be the other. His interest in outside ideas was in part rebellion against the society in which he had been raised. He was prone to parody and iconoclasm, and his notorious club of cronies, the All-Joking, All-Drunken Synod of Fools and Jesters, openly and often cruelly mocked Russian institutions from the church to traditional manners. There is certainly no likelihood that he saw modernization in philosophical terms: this was the man who left his own name carved as graffiti into the doorsill of religious reformer Martin Luther's home. Instead, it was intensely practical. Peter appreciated that a backward Russia was a weak one, and that a weak state was a vulnerable one.

He addressed the immediate needs of administration,

tightening the grip of the state on nobility, church and peasantry alike. He was ruthless when challenged; he had his eldest son tortured under suspicion of plotting against him, a torment from which he died. He made of Russia a great power, forcing the West to pay attention to what had hitherto been considered a "rude and barbarous kingdom" (as sixteenth-century English navigator Richard Chancellor had dubbed it). This was reform only insofar as it was needed for security, though, and never sought to be anything more. It was modernization as envisaged by the soldier and carpenter, the shipwright and amateur dentist. It would take another "Great"—Catherine—to tackle modernization of the mind and the soul.

Further reading: Although Robert K. Massie's very readable *Peter the Great: His Life and World* (Knopf, 1980) is often considered the gold standard, I would suggest that *Peter the Great* by Lindsey Hughes (Yale University Press, 2002) is the best of the many biographies of the man: judicious, well-written and with just the right mix of skepticism and respect. Foy de la Neuville's *A Curious and New Account of Muscovy in the Year 1689* (SSEES, 1994) is one of the more interesting (if not always accurate) primary accounts of the time, written by a Polish envoy, and the text also has the virtue of being freely available online. Considering how far warfare defined Peter's reign, Peter Englund's *The Battle That Shook Europe: Poltava and the Birth of the Russian Empire* (I.B. Tauris, 2013) does a good job of dissecting this particular, crucial engagement and exploring its wider context.

5

"I SHALL BE AN AUTOCRAT: THAT'S MY TRADE"

Timeline

AN IMPERIAL   STRIDE!

Russia.   Constantinople.

EUROPEAN POWERS.

© BRITISH MUSEUM

An Imperial Stride, *a satirical English cartoon (1791)*

English cartoonists took no prisoners when they lampooned the public figures of the day, but while not an especially flattering representation of Russian Empress Catherine II—Catherine the Great—it is in many ways a mark of Russia's new status. In a dig at her ambitions to challenge the Ottoman Empire, here she is striding in one step from Russia to Constantinople, while below her the spiritual and secular lords of Europe express their ribald admiration. "Never saw any thing like it," says Louis XVI of France, and "What a prodigious expansion," swoons England's George III. The Ottoman Sultan sighs that "The whole Turkish Army wouldn't satisfy her."

Cheap shots at her (largely mythical) sexual appetites aside, what is noteworthy is firstly that Catherine is represented not as some exotic Asiatic monarch, but a wholly European one, a culmination of the process begun by Peter to bring Russia back into the West. Secondly, while Russia never did manage to take Constantinople, the thought of it doing so was considered entirely possible. Russia was no longer of negligible significance and distant bearing, but part of the fractious European family of nations.

After all, Catherine the Great (r. 1762–96) did more than just shape Russia's eighteenth century; she also shaped its image and place in the world. In that, she was in many ways what today we might call a mistress of spin. She expanded and blinged out the Winter Palace in St. Petersburg to rival the grandeur of the French monarchs' Versailles. She kept up a spirited correspondence with the philosophers of the day, notably Voltaire, even while holding them safely at arm's length so they could not see how many of her claims about Russia were simple hype. At a time when the country was being squeezed to support war against the Ottomans, she told Voltaire that "our taxes are so low, there is not a peasant in Russia who does not eat chicken when he pleases." She was indeed a reformer, and sought to bring culture and literacy, progressive policies and sound laws to the country. Catherine was the very image of that eighteenth-century European ideal, the "enlightened despot," dragging countries into the future by the power they inherited from the past.

Arguably, though, the more she tried to turn Russia into a European nation, the more she brought forth some unavoidable contradictions in the design. Catherine's golden era was in many ways to prove nothing but gilded brass, a thin patina of European culture over a nation that was increasingly falling behind and away from the courts, factories, shipyards and universities of the West. One of her favorites, Grigory Potemkin, was said to have built fake villages to give an impression of cheery plenty for a visiting Catherine. This is probably something of a fable, but ultimately Catherine's Enlightenment Russia was a "Potemkin nation," trying too hard to persuade everyone else—and itself—that it was something it was not. It would take an upstart Corsican artilleryman truly to puncture many of the myths of eighteenth-century Russia, when Napoleon smashed across Europe and into its lands.

Fundamentally, after all, the country was still mired in the Middle Ages socially and economically. The overwhelming majority of the population were peasants; mostly serfs working on land owned by the state, aristocrats or the church. This would scarcely change over the century: 97 percent lived on the land in 1724, 96 percent in 1796. Serfs were chattels, who could be sold or transferred around the country as family units, and had no claim to the land they worked. Although some token attempts had been made to introduce Western farming methods, these had not accounted for much, sometimes because the heavy soils and hard climate did not allow, but often because of a lack of skills, training, investment capital and interest, so agricultural productivity was still close to medieval levels.

Domestic and international trade did grow, especially as Russia gained ports on the Baltic and Black Seas, and with it a mercantile class began to emerge, but it was very small. Peasant traders handled much small-scale domestic commerce, and foreigners and noblemen much of the rest. The state would be in a permanent financial crisis through the century, tax collection never keeping up with expenditures on wars, prestige building projects and the court, and having to fill the gap with promissory notes and printing money. While the state bureaucrats and richer nobility were literate and becoming exposed also to foreign ways, many of the rural gentry could often neither read nor write. This was hardly the country Catherine the Great would sell to the West.

A Time of Empresses

Russia, so traditionally chauvinist, was about to have to get used to women being on the throne. When Peter died in 1725, he had on the one hand established the principle that a tsar could name his successor (from his family), but on the other failed actually to nominate anyone. He had previously declared his second wife, Catherine, as tsarina, empress, but it is questionable whether she could have taken power simply on that basis. Instead, she was considered a suitable figurehead by a cabal of figures who had risen under Peter, led by the shrewd but deeply corrupt Prince Alexander Menshikov. Calling on the Guard regiments—who not for the first time would prove kingmakers, or in this case empress-makers—they installed her as Catherine I (r. 1725–7) in a virtual coup d'état, fearing that otherwise traditionalists from the older boyar families would simply return to power.

She had two daughters of her own, but Russia was not ready for a matrilineal descent, and so Catherine had to assent to the naming of the only male-line grandson of Peter I as her heir. When she died in 1727, the 12-year-old Peter II (r. 1727–30) was duly crowned, the ubiquitous Prince Menshikov as his regent. Fate, though, seemed disinclined to let this sexism pass, and he died a mere three years later, leaving no male heirs. The next in line would have to be found from the children of Ivan V, Peter's co-tsar, and this meant either the eldest, Catherine, or her younger sister, Anna. Like it or not, the Russians were going to have another empress.

While Catherine was older, she was married to a German, Karl Leopold of Mecklenburg-Schwerin, whom the boyars feared would seek to exert influence over Russia if his wife became tsarina. Instead, the Supreme Privy Council, the body that had replaced the Boyar

Duma, opted for the widowed Anna. As with Catherine, though, the intention was that she was to be a figure-head. Prince Dmitry Golitsyn, chairman of the Council, presented her with a set of "Conditions" that she was expected to adopt. They discovered it was easier to demand obedience than to enforce it. Once crowned, Anna (r. 1730–40) tore up the "Conditions" and dismissed the members of the Council, filling it with more agreeable candidates. Golitsyn would end up dying in prison, and while he has since been celebrated as a man who tried to bring constitutional rule to Russia, one can wonder whether this was really out of principle or simply because he thought this was his chance to be the power behind the throne.

Ten years later, close to death, Anna made her two-month-old grandnephew, another Ivan, her heir, appointing her German lover, Ernst Biron, regent. This was an attempt to secure both Ivan V's bloodline and also Biron's future. She had never been a popular figure, though, and her tendency of filling the court with German relatives and cronies had alienated populace and boyars alike. Ivan VI (r. 1740–1) was crowned, but within three weeks Biron had been banished to Siberia, and just 13 months after his coronation, the unlucky child–tsar and his family would be imprisoned in a fortress in Russian-controlled Latvia after a coup by Elizabeth, daughter of Peter. Ivan V and Peter had managed to coexist as monarchs, but their bloodlines were seemingly locked in war.

Energetic, intelligent and charming, Elizabeth had won over the elite Preobrazhensky Regiment, and in 1741, they seized the Winter Palace, Ivan and the throne for her in one bloodless night. The 33-year-old Empress Elizabeth (r. 1741–62) ushered in an era of elegance, extravagance and diplomacy. Russia became ever more present as a major European power. A war with Sweden was ended, with Russia taking southern Finland, while it became a key player in the Seven Years War (1756–63) against a rising Prussia. In 1762, Frederick the Great of Prussia was on the verge of defeat when news came that Elizabeth had died. Childless, and desperate to keep Peter the Great's bloodline on the throne, the only heir she could choose was her nephew, Peter of Holstein-Gottorp. German-born, even though Elizabeth had tried to ensure he had a Russian education, scarred by smallpox and obsessed with toy soldiers, Peter III (r. 1762) would only reign for 186 days. However, his elevation crucially opened the door to the Winter Palace to his wife, Princess Sophie of Anhalt-Zerbst, who would be known in history as Empress Catherine the Great.

From Sophie to Catherine

Sophie Friederike Auguste von Anhalt-Zerbst-Dornburg was born to Prussian German aristocratic stock of considerable connections but relatively little fortune. As was the fate of girls in such circumstances, the expectation was that she would be married off for the good of the family, regardless of her own inclinations. Certainly the choice of her second cousin, Peter of Holstein-Gottorp, owed everything to politics and nothing to affection. Peter's aunt, Tsarina Elizabeth, was eager to build ties to Prussia, and Sophie's ambitious, manipulative mother was enthused by the prospect of a daughter on the Russian throne and a chance to spy for Frederick II of Prussia (she was eventually banned from the country for that very reason).

At age 15, Sophie traveled to Russia. Peter she found obnoxious, but she neither had much choice in the matter nor was she unaware of the opportunities for an otherwise impecunious young Prussian princess. It hardly hurt that Elizabeth clearly took a shine to her. So in characteristically enthusiastic style, Sophie threw herself into learning Russian, was baptized into the Orthodox faith with the Russian name Yekaterina— Catherine—and married Peter in 1745.

Peter did not improve on closer acquaintance and the two of them lived largely separate lives, taking lovers and pursuing their own interests. Peter loved to play with both toy soldiers and real ones, putting his servants through an elaborate and demanding drill of a morning. The vivacious and shrewd Catherine, by contrast, actively courted the all-important Guard regiments. When Peter III ascended the throne in 1762, he managed in short order to make himself even more unpopular, not least by prematurely withdrawing from the war with Prussia (he was a great admirer of Frederick the Great, even referring to him as "my master").

The irony was that the German Tsarina Catherine seemed more loyal than the Russian-blood Tsar Peter. Having endured 17 years of marriage to the man, she was clearly prepared to seize the opportunity to rid herself of him, though, and take his place. While he was spending time with his relatives in his country estate, Catherine was back in St. Petersburg, plotting. Resplendent in a guards' uniform, she visited the Izmailovsky and Semyonovsky regiments and appealed to them for their support. She had the church onside, she had key figures within the government onside, and she had the Guards. Peter was arrested and forced to abdicate, and shortly thereafter was killed, and Catherine II (r. 1762–96) was empress.

As usual, history, tradition and ritual were scoured to justify political pragmatism. Fortunately, a tenuous path could be tracked through Catherine's genealogy to the Ryurikid dynasty, and Catherine I's example of succeeding Peter the Great, however questionable at the time, now became considered precedent. Although there would be occasional word of conspiracies—and more serious threats from risings in the countryside—

Catherine would remain firmly in power until her death.

Educated, intelligent and coming from the more cosmopolitan European nobility, she was committed to reform as a means of raising Russia to Western levels. If Peter the Great's real focus had been military, and the state-building measures this demanded, hers was cultural and intellectual, and the programs that flowed from this. In Europe, this was the era of "enlightened despotism," of autocratic rulers who claimed to be inspired by the Enlightenment values of reason, scientific process, freedom and tolerance, and to rule in the interests of their people. Often, this actually proved to be more about despotism than enlightenment, and Catherine herself famously said, "I shall be an autocrat: that's my trade."

Nonetheless, she also had a clear enthusiasm for presenting herself and her country as a European power, at the forefront of the Enlightenment. Western fashions and culture became de rigueur at her cripplingly lavish parties and celebrations (by 1795, over an eighth of the total state budget went on court expenditures). Just as hiring Dutch and English shipwrights did not modernize Russia overnight, though, nor did corresponding with European philosophers and buying Western art collections reshape the nation. To be honest, her reputation was as much not about what she did, but what she wrote, and what others wrote about her.

Empty Enlightenment?

That did not mean Catherine did nothing. Quite the opposite: this was a time of considerable progress and change. Foreign books once banned or ignored were translated, and vaccinations against smallpox introduced over the complaints of many. She espoused religious tolerance (while at the same time seizing the last of the church lands) and ended the use of torture (in theory). Although grandiose plans to provide universal education predictably came to nothing—not least because peasants could see little point in it—her reign did see an expansion of schools and universities, with even some women being admitted.

Yet there was a void at the heart of her programs. She seemed genuinely to believe in the importance of liberty and law, which to be meaningful have to constrain the state and the monarch. But she was also an unabashed autocrat, unwilling to brook protest or dissent. Was she serious about legality and reform, or just playing a role? In 1766, for example, she called a Legislative Commission that was to be made up of representatives of the aristocracy, gentry, townspeople, state peasants and Cossacks—not the serfs—to consider a new code of laws. They were presented with Catherine's *Nakaz*, an "Instruction," being a 22-chapter statement of the principles she wanted embodied in this code. She had worked on this document for almost two years and although she heavily and often word-for-word cribbed from the French philosopher Montesquieu, the Italian jurist Cesare Beccaria and other European thinkers, it nonetheless is an impressive and progressive treatise, which marries a continued commitment to absolutism with liberal notions of equality and legalism.

However, when the commission met next year, it became clear that this haphazard collection of boyars and burghers, soldiers and squires, clerks and Cossacks, was unqualified, disunited and unsure of its mandate. It met 203 times and discussed everything from nobles' privileges to merchants' rights, but failed to reach any conclusions or make a single recommendation. Eventually, after the Russo-Turkish War broke out in 1768, it was suspended and never recalled.

But does that mean it was wholly pointless, nothing more than an exercise in cosmetic constitutionalism? Not quite. First of all, Catherine cannot be wholly blamed for its impotence. This was an experiment in consultation, and as much as anything else an opportunity precisely to see if there was any consensus in the country. (There wasn't.) It also gave her new insights into the priorities and concerns of social groups that otherwise rarely got to have their views heard, views that did make their way indirectly into much future legislation. It also brought in representatives of the petty rural gentry, who often didn't stray far from their estates except when called upon in war. This reminded the boyars (the term persisted for the more powerful aristocrats) that there were alternative power bases to which Catherine could appeal.

Power and Purpose

After all, behind all the grandiose rhetoric about egalitarianism was a crucial renegotiation of power between the state and an aristocracy that still dominated the army, the civil service and the countryside. They were not investors with wealth in stocks and shares, nor were all in government service. Rather, their wealth was still based on land, and the peasants and serfs who worked it, and so serfs had to remain serfs. As empress, Catherine herself owned half a million of them, and the state another 2.8 million. She gave them more rights to petition local governors if their masters abused them, but meanwhile she also gave the landowners the right to exile serfs to Siberia; win some, arguably lose more.

In his blink-and-it's-over reign, Peter III had issued a barrage of new laws and decrees, including the "Manifesto of the Freedom of the Nobility" that further ate into the service obligations of the nobility. In its words, "no Russian nobleman will ever be forced to serve against his will; nor will any of Our administrative departments make use of them except in emergency cases and then only if We personally should summon them." A cynic might suggest a weak man in a weak position had been trying to buy himself the support of the Russian aristocracy. In any case, Catherine— presumably also mindful that a monarch elevated by a coup could just as easily be deposed by another—continued this dismantling of Peter's service state.

In 1785, she went further, issuing her "Charter to the Gentry." This confirmed exemptions from compulsory state service and taxation, sweeping rights over serfs and full hereditary rights to all estates. The gentry were also granted the right to establish their own assembly in each *guberniya*, or province. In many ways this was classic Catherine: conscription in the guide of concession. She clearly had come to realize that one of the problems with the empire was precisely that it was too focused on a single monarch. She wanted not to weaken autocracy but to make it more responsive and thus stronger by creating intermediary institutions that would handle much day-to-day governance rather than require everything to be run either by St. Petersburg or through individually appointed governors with all their temptations to corruption and sloth. That same year, after all, she also granted towns and cities their own charters, building structures of local governance there.

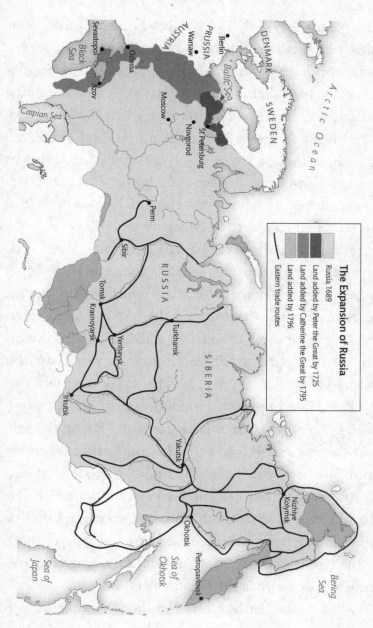

The Expansion of Russia

Russia 1689
Land added by Peter the Great by 1725
Land added by Catherine the Great by 1795
Land added by 1796
Eastern trade routes

In many ways, this is what was important about Catherine's era of reforms. Not the vain and vapid correspondence, nor the hollow pledges of Enlightenment values. Catherine may have disliked the death penalty, but she turned a blind eye when the brother of one of her favorites murdered Paul III, and Yemelyan Pugachev, the leader of the largest peasant revolt in Russian history—and who, evoking the False Dmitries of previous eras, claimed to be Paul III—was beheaded and chopped to pieces in Moscow in 1775.

Her foreign policy was equally pragmatic, even if always cloaked in the right, almost apologetic rhetoric: "I have no way to defend my borders but to expand them," she claimed. She certainly did that, and Russia's territories grew by more than half a million square kilometers during her reign. She aggressively warred against Poland–Lithuania, taking part in the three partitions that saw Russia end up with Lithuania and most of eastern Poland. The Ottomans were a particular target, though, as Catherine saw Russia's greatest opportunities to the south. She never did make that "imperial stride" to Constantinople, but defeated the Turks in the wars of 1768–74 and 1787–92. As a result, she took southern Ukraine and, in a move that would have historical repercussions for the twenty-first century, annexed the Ottoman dependency of Crimea for Russia in 1783.

At heart, the "enlightened despot" Catherine was more despot than enlightened. Her words in the *Nakaz* were unambiguous: "The Sovereign is absolute; for there is no other Authority but that which centres in his single Person, that can act with a Vigour proportionate to the Extent of such a vast Dominion ... Every other Form of Government whatsoever would not only have been prejudicial to Russia, but would even have proved its entire Ruin." But she was a smart despot, who understood that the old ways of rule in Russia were increasingly out-of-date. As the ever-quotable empress said, "A great wind is blowing, and that gives you either imagination or a headache." She undoubtedly had not come up with the answers as to how Russia could change—but she was beginning to ask the question.

After Catherine

Catherine died a natural death in 1796 and was succeeded by her son Paul I (r. 1796–1801). Gossip had questioned whether he was truly Peter III's son, and Catherine had had little time for him as he grew up. In fact, she seriously considered bypassing him in the succession altogether and declaring his son Alexander her heir. His reign was overshadowed by Catherine's personality and legend. In what might seem his own form of adolescent rebellion—although he was 42 years old at the time of his coronation—he bucked against an education that was designed to mold him into the perfect Enlightenment reformer and instead espoused an inflexible and authoritarian conservatism.

He hurriedly issued the so-called Pauline Laws, establishing that henceforth the throne would only go to the next male heir in line: no more empresses, no more dangers of someone finding their own son made successor in their place. He made no secret of his disdain for the aristocracy, showering riches on a handful of cronies, and treating the rest with contempt. Like his father he was passionate about the military, but also like his father he had little understanding of real generalship and again indulged in parades and detailed decisions about new uniforms.

The shape of Russia's military would matter, though. This was the era of the French Revolution, and being both a convinced autocrat and of mystic bent (in 1798, he was elected Grand Master of the Maltese Order of the Knights of St. John of Jerusalem), he thought of this in terms of a crusade against anarchy. In 1799, Russia joined Austria, Turkey, Britain and Naples in declaring war on France. The coalition would fall apart, and once Napoleon had made himself "First Consul" in 1799, Paul began talking of a possible alliance with France against the Ottomans. He even drew up plans to send a force off in the direction of British India: to Russia's aristocratic elite, it looked as if he wanted to take on the world.

He had become dangerous, in a dangerous time, and in the eyes of dangerous men. In 1801, a gang of dismissed officers burst into his bedchamber and tried to make him sign an abdication decree. When he resisted, he was strangled. Did his eldest son, the 23-year-old Alexander, know this was going to happen? All we do know is that he never punished the assassins. Ultimately, though, we know much about what Tsar Alexander I (r. 1801–25) did and said, but it is frustratingly hard to come to terms with who he was. A liberal? A conservative? One of his own mentors, Mikhail Speransky, called him "too weak to rule, and too strong to be ruled." But in many ways maybe it didn't matter, because his reign would be overtaken by a war with France that would see Moscow burned, Russian soldiers in Paris, and the world changed.

The country would finally have to come face-to-face with the challenge of reform. Peter the Great had tried to force modernization from above, and had made some progress but not enough. Catherine the Great had tried to inspire modernization from above, and had made some progress, but not enough. It was to become clear that real change would have to come from below, and this would prove a terrifying prospect for some, an exhilarating opportunity for others. The opening words of the first chapter of Catherine's *Nakaz* were that "Russia is a European state." But in one of her letters to the French writer Denis Diderot, she wrote that "you philosophers are lucky men. You write on paper and paper is patient. Unfortunate Empress that I am, I write on the susceptible skins of living beings." It was time to see if Russians could not simply be decked out in European finery, or taught to read European books and admire European art. Instead, could they have a European identity—defined in terms of culture and values as much as technology and tradecraft—written on and within their skins?

Never mind the stories Russia told the rest of the world, epitomized in Catherine's *belles lettres* to the philosophers of the West. The real question would be what stories Russians would tell themselves about themselves. Peter and Catherine had sought to create narratives that placed Russia in Europe, without necessarily truly thinking through what that meant. They had also told that story to foreigners and to the elites. The slow spread of education and literacy, the emergence of a middle class often looking for its place in the world, influences from the French Revolution to Marxism, all of these would also play their part in ensuring that the nineteenth century was one in which Russia's identity would be contested more vigorously, and by more players, than ever before.

Further reading: Robert K. Massie's *Catherine the Great* (Head of Zeus, 2012) is a massive, benchmark biography, although Simon Sebag-Montefiore's *Catherine the Great and Potemkin: The Imperial Love Affair* (Weidenfeld & Nicolson, 2016) is a more fun read. If you do want to read her correspondence, *Catherine the Great: Selected Letters* (Oxford University Press, 2018), translated by Andrew Kahn and Kelsey Rubin-Detlev, is a good collection, and there are always the *Memoirs of Catherine the Great* (Modern Library, 2006).

6

"ORTHODOXY. AUTOCRACY. NATIONALITY"

Timeline

© MARK GALEOTTI

The Cathedral of Christ the Saviour, Moscow

Within view of the Kremlin, the white-walled and gilt-domed Cathedral of Christ the Saviour was originally conceived by Alexander I to celebrate Russia's victory over Napoleon. It was to be a neoclassical structure, reflecting the style dominant in the West at the time. The original site proved unsuitable, though, so his son, Nicholas I, decreed that the cathedral would instead be built where it is now, but he favored a more traditional style, evoking Russian traditions and the glories of the Hagia Sophia in Constantinople. The outer structure was completed under Tsar Alexander II, who largely ignored the project, and the cathedral was finally consecrated in 1883, on the eve of the coronation of Alexander III. It would later be dynamited under Stalin and rebuilt in the 1990s, largely to Nicholas's original design.

In granite, marble and 20 tons of gold, the church is a metaphor for changes in politics and priorities over the years. Alexander I wanted to show that Russia was rich enough and European enough to be able to build a towering edifice that matched the tastes of the times. Nicholas I wanted to demonstrate that Russia didn't need to keep up with the neighbors, and could cleave to traditional styles and imperial aesthetics. Alexander II was busy and not especially concerned with churches, but rather factories, courthouses and schools.

Should Russia simply try to look like the equal of a Western power, without actually trying to be one? Should it stick to its own ways? Should it try and grasp the essence of modernization, rather than just the appearances? Napoleon's invasion of 1812 was defeated by logistics and demographics, but Russia convinced itself that its ability to weather his advance and join the counteroffensive proved that it was stronger than everyone had assumed. This was a perfect excuse to justify putting off the social, political and economic modernization that Russia desperately needed. Any reforms would inevitably bring with them uncertainty and unrest, after all, as Alexander II's reign was to demonstrate.

The nineteenth century was, therefore, one of competing myths, each of which directly bound Russia with Europe. To reformers, it needed to be more Western. To conservatives, Russia needed to reject the West, lest chaos be unleashed. Meanwhile, revolutionaries were increasingly looking to ideologies being elaborated in Europe and seeing in them magical solutions that could somehow leapfrog Russia into the forefront of the socially and economically advanced nations of the West. Unwilling to accept the changes reshaping Europe, unwilling to accept exclusion from Europe, Russia was being torn apart by the contradictions in the stories it told itself about itself.

General Winter and Mother Russia

Much of Russia's nineteenth century would be defined by Napoleon's decision to invade in 1812. The war that followed—called the "Patriotic War" by the Russians—was a terrible one for them, and consequently the victory that followed was a transformative one. Alexander I (r. 1801–25), the tsar whom one of his contemporaries said "knocked at every door, so to speak, not knowing his own mind," had first turned against France, then with it, in an alliance of inconvenience that had foundered in 1810, as Napoleon failed to keep his promises to help Russia against the Ottomans. The French emperor was a man who could not stop for either personal or political reasons, so in 1812, in a fateful act of hubris, he led his *Grande Armée*, the largest expeditionary force the world had ever seen, into Russia. It was time, Napoleon purportedly said, "once and for all to finish off these barbarians of the North," who "must be pushed back into their ice, so that for the next 25 years they no longer come to busy themselves with the affairs of civilized Europe."

Napoleon's army of French veterans, Polish lancers, Austrian riflemen and Piedmontese sharpshooters would be broken in Russia. In part this was down to the stubborn resistance of the Russians but also thanks not only to the country's traditional ally, "General Winter," but the sheer size of Mother Russia's bounds. Faced with an enemy able and willing to take fullest advantage of strategic depth and withdraw before his advance, burning crops and spoiling wells on the way, Napoleon became increasingly infuriated that the Russians simply weren't willing to play by his rules. Finally, at Borodino, the Russians stood and fought, losing perhaps a third of their army to withering barrages of cannon fire in the bloodiest day of fighting of the Napoleonic Wars. Yet they retreated in good order. The Russians even abandoned Moscow, but refused to surrender.

Napoleon brooded in Moscow for a whole month, sure that the Russians would sue for peace. When they did not, as the stocks of food for his men and fodder for his horses ran lower and lower, he was forced to retreat. Harried by Cossack hit-and-run raids, ambushed by angry peasants, ravaged by hunger and disease, the

Grande Armée shrunk with every pace of its retreat, while Napoleon spurred ahead of them to Paris, desperate to secure his position and spin the defeat as a victory. Of the 685,000 men who marched into Russia, only some 23,000 stumbled out alive.

The war was not over, but momentum had swung against Napoleon. Seeing the opportunity for victory, Prussia and Austria joined Russia in rolling westward, and the Duke of Wellington led his British, Spanish and Portuguese forces across the Pyrenees into France. In 1814, Napoleon abdicated and left for exile in Elba (from which he would briefly return in 1815, but that's another story), France was sheared of 20 years of gains, the tsar could add portions of Finland and Poland to his empire, and Russian officers could toast their victory on the Champs-Élysées.

After the Antichrist

Before the war, the ever-inconstant Alexander had first toyed with reform, even daring to propose candidates for higher government positions ought to sit exams, before falling under the influence of Count Alexei Arakcheyev. A ruthless zealot of predictably bad temper and unpredictably vicious whimsy—he would weep at the beauty of nightingales' song and then have all the cats in the neighborhood killed so they did not prey on them—Arakcheyev encouraged Alexander to embrace the mystical messianism of his father. The

French Revolution had unleashed a wave of radicalism in Europe and in due course the rise of Napoleon. In Alexander's mind the consequent anarchy threatening Europe came to seem an almost satanic threat (he became fixated by the news that a Russian scholar had proved through cabalistic numerological analysis that the letters in the phrase "L'Empereur Napoléon" added up to 666, the "Number of the Beast"), and victory a triumph of righteous order.

It also seemed to reaffirm the strength of the Russian state. The eighteenth century had been dominated by concerns about how to modernize, how to catch up with the West—and how dangerous this might be for the existing order. Would the rise of "new men" within the government apparatus supplant the aristocracy? Would the spread of education further revolutionary sentiment? Victory over Napoleon became a convenient myth of the system's fundamental health. After Borodino, Napoleon wrote that "the French showed themselves to be worthy of victory, but the Russians showed themselves worthy of being invincible." Arguably this was a poisonous bequest to his conquerors: dangerous is the day any regime tries to convince itself of its own invincibility.

Doubly so when some of its best and brightest are drawing the very opposite conclusions. Since Catherine's day, French language, literature and ideas had been considered the very peak of sophistication. Many young officers, drawn from the educated elite, had been enthused by the ideals of the revolutionary age, and then exposed to them in France. The early years of Alexander's reign had generated hopes that change was coming to Russia, hopes the subsequent conservative reaction had dashed. Secret societies, radical factions and conspiratorial movements bubbled away beneath the orderly surface of the regime, some advocating constitutional monarchy, others outright republicanism. By the 1820s, they had found common cause in concluding that their only hope of change was in a violent coup. They even set a date, for December 1825.

Then Alexander had the poor grace to die, of typhus, just under a month before they were ready to strike.

The Soldier Tsar

Why let a good plot go to waste, though? They decided still to go ahead with their plan, just as the new tsar Nicholas I (r. 1825–55) was being crowned, and in the process ensure that his reign would be devoted to an undeviating defense of the status quo. Russia's story is full of such tragic ironies.

Nicholas had not even been meant to be tsar. Alexander left no legitimate heirs but two brothers. The younger, Nicholas—Nikolai—had instead trained as a soldier because Grand Duke Konstantin, governor of Poland, was next in line. But Konstantin married a Catholic Polish countess in 1820, and thus had secretly surrendered his claim to the throne. Dutiful, energetic and unimaginative, Nicholas then accepted the crown just as the liberal conspirators rushed to try and seize the initiative. What became known as the Decembrist Revolt of 1825 saw some 3,000 young army officers and their fellow travelers take to the streets in St. Petersburg. They demanded a constitution, they wanted reform, they got Nicholas. Cannon blasted them out of Senate Square, and loyal troops rounded the survivors up at bayonet-point, most ending up exiled to Siberia. The new tsar would come to power in battle, and from the first saw his role as being a soldier in the frontline against anarchy and insurrection. He saw no contradiction between his view that an autocrat had to be "gentle, courteous and just," and building a police state and ruthlessly suppressing liberal writers and radical thinkers alike.

He looked at the empire as he would at an army. Just as an army needed discipline, so too did a nation, and at a time of revolutionary sentiment, it needed something to hold it together. In the wake of the Decembrist Revolt, which was considered in part a product of dangerous, foreign-inspired freethinking among the educated young, it fell on Count Sergei Uvarov, as education minister, to provide an answer. In 1833, he proposed the notion of "Official Nationality," a doctrine that claimed that Russia's traditional values needed to be defended against alien notions. The formula that was adopted was "Orthodoxy. Autocracy. Nationality."

Leaning on Russia's role as the last true Orthodox
state was nothing new, but after a century of lip service
to the idea of getting beyond tradition and embracing Western rationalism, this became a justification for
quite the opposite. "Europe" was now a source of con-
tamination that Nicholas said was "in harmony with
neither the character nor the feelings of the Russian
nation." God meant Russia to be Russia, not a pallid
copy of Western Europe. This was the job of autocracy, undiluted by constitutionalism. This did not mean
tyranny for its own sake but what one could call, in
contrast with Catherine's ideal, unenlightened despotism: rigid central power in the name of the common
interest. In the pursuit of national unity and a single
loyalty, all subjects of the tsar should embrace a single
faith and values. This was an era of "Russification," as
the Catholic Church in Poland came under new restric-
tions and Poles, Ukrainians, Lithuanians and Bessara-
bians were pressed to learn Russian.

Astolphe-Louis-Léonor, Marquis de Custine, was a
liberal French aristocrat who traveled through a Russia that clearly appalled him, and he opined that "This
empire, vast as it is, is only a prison to which the emperor holds the key." The irony was that the tsar who
has largely gone down in history as the epitome of
the unreasoning martinet, the humorless defender of
a dying and despotic order, was actually deeply skep-
tical of it. He genuinely believed in his divine right to

rule, but believed that this required him to be hard-working, mindful of his responsibilities to Russia and the divine. He created the Special Corps of Gendarmes and the fearsome political police of the Third Section of His Imperial Majesty's Own Chancellery, both under General Count Alexander Benckendorff, but genuinely saw them as protectors of the people, policing them for their own good. Nicholas notoriously gave Benckendorff a handkerchief with the enjoinder to wipe away the tears of his subjects. He also oversaw a regime of censorship that often verged on the farcical—in some cookery books, references to "free air" were excised as sounding too subversive, and both Alexander Pushkin and Fyodor Dostoevsky, greats of the Russian literary canon, would fall foul of the Third Section—but which was nonetheless seen as essential to hold back the tide of destructive ideas from the West.

He was not a self-indulgent tsar, and over time became increasingly disenchanted with the triviality and pomp of court entertainments. He certainly had no illusions about the aristocracy: according to legend, he told his ten-year-old son Alexander, "I believe you and I are the only people in Russia who don't steal." Indeed, his reign was noteworthy for the proliferation of generals appointed to ministries and Baltic German aristocrats (like Benckendorff) from the empire's north-western margins elevated to key positions: he felt he had to go outside the usual pool of officials and noblemen in the hope of finding honest and efficient personnel. Sadly, that often didn't work, either.

Most strikingly, Nicholas also disapproved of the institution of serfdom. Through his reign, he would convene a series of secret commissions to try and find some way of squaring the circle: how to abolish a system of land-slavery that was inefficient, inhumane and a source of periodic uprisings without totally destabilizing the whole social order and alienating the rural gentry, the landlords who were the backbone of the tsarist order in the countryside. Nicholas was brave enough faced with physical danger, but he never dared tackle this challenge, concluding that "There is no doubt that serfdom in its present position is evil … but trying to extinguish it now would be a matter of even more disastrous evil." And why take the risk? Hadn't victory over Napoleon proved that, however backward it might seem, the Russian system was strong enough to triumph and survive? So Russia's rulers told themselves, as long as they could.

The Gendarme of Europe

For a while, they could continue to do so. For most of his reign, Nicholas was a successful warrior-tsar. He certainly devoted passion, time and excessive amounts of money to the Russian military. His army would grow to a million men—out of a total population of 60 to 70 million—but it would later become clear that he mistook spit and polish for real combat effectiveness.

Like Alexander, Nicholas considered championing the traditional order as an internationalist duty. During his reign Russia would become known as the "Gendarme of Europe" for his enthusiasm to help fellow monarchs stamp out the fires and embers of revolution. In 1831, his army crushed a revolt in Poland triggered by his rolling back of the Poles' constitutional rights. Once a subject kingdom with its own parliament, Poland was reduced to the status of a mere province, under an appointed governor. When a series of revolutions erupted across Europe in 1848, even though Russia was suffering from famine caused by unusually poor harvests and a cholera epidemic, his troops would again march in the name of the status quo. Having already helped the Austrians suppress the revolt of the Free City of Krakow in 1846, Nicholas broke the 1848 Moldavian national movement and then sent his armies to join the Habsburg Empire in putting down the Hungarian revolution in 1849.

Again, the double-headed eagle looked both ways. Nicholas was at once committed to—as he saw it—saving Europe from its own ungodly and illegitimate dalliance with liberalism, as well as protecting Russia from European ideas. Aware of the advances in Western science and technology, he wanted to adopt the elements of the West that looked useful, while ignoring the social, political and legal contexts from which they sprang. Without a thriving mercantile class to generate investment capital, without free and open debate in universities and educated circles to generate ideas, and without greater social mobility to generate new cohorts of innovators and skeptics, Russia would always remain backward, desperately trying to adopt and adapt the inventions of others.

This did not matter so much while Nicholas's armies were deployed against rioters, even when their enemies were Persians and Turks. It would matter a great deal when they found themselves fighting the British and the French—the most advanced military powers of the age—in Crimea, a war Nicholas had never wanted to fight, and which was paradoxically triggered by his desire to protect the European status quo.

Despite Western portrayals that presented him as just one more incarnation of a Russia bear eager to gobble up territories to the south and the southwest, in the main he felt he was acting to maintain stability. This was a real challenge when it came to Russia's old rival, the Ottoman Empire. There was traditional bad blood there, not least given the Muslim Turks' occupation of Orthodox Christian lands (and the "Second Rome," Constantinople). Nicholas feared, though, that any serious pressure on this decaying empire risked bringing it down, causing chaos in southeastern Europe, angering Ottoman allies France and Britain and alienating ally Austria. Instead, he wanted to keep the Ottomans weak enough not to be a threat—and potentially even becoming Russian vassals—but not so weak that their fractious empire broke apart. Nicholas also needed to guarantee passage rights through the straits of the Dardanelles and the Bosphorus, as this was a crucial trade route for Russia, especially its grain exports.

The Greeks had been battling for independence since 1821, and in 1827, fearing an Ottoman collapse or a unilateral Russian intervention, the British and French joined with the Russians to force the empire to grant them at least autonomy. At Navarino, the allied fleet decisively defeated a larger but antiquated Ottoman force, but Sultan Mahmud II was still not willing to concede. He closed the Dardanelles to Russian shipping. In response, Nicholas put 100,000 men into the field and, after some hard fighting, by 1829 the Ottomans had been forced to sue for peace.

Crimea and Punishment

Ultimately, though, the "Eastern Question" was less about Turkey than European great-power politics. Britain feared Russian expansionism: both London and St. Petersburg saw fiendish long-term strategy in their rival's clumsy short-term responses to a chaotic world. France's newly minted emperor, Napoleon III, was looking for glory. The Ottomans feared Russia. And Nicholas not only feared being locked away from the

Mediterranean, but resented what he saw as Western double standards. He found his sentiments accurately reflected in a report by a Russian scholar, Mikhail Pogodin, who wrote that "we can expect nothing from the West but blind hatred and malice."

A squabble over the rights of Christians in the Ottoman-occupied Holy Land led to Nicholas asserting himself as guardian of the Orthodox community. Attempts to broker an agreement fell through, and in 1853 the Ottomans—believing they had British and French support—declared war on Russia. It started badly for them, as Russian troops crossed the Danube into Romania and Russian ships smashed a naval squadron at Sinope. Fearing an Ottoman collapse, the French and British rushed forces to the Balkans, just as the Russians pulled back.

Having stirred up jingoistic sentiment at home, neither government could afford to remain, as Karl Marx and Friedrich Engels put it, with "the French doing nothing and the British helping them as fast as possible." So they chose instead to turn their attentions to the Crimean Peninsula and Russia's main base on the Black Sea, Sevastopol. It took almost a year for an allied force to take the city in a war marked as much for the incompetence as the bravery shown on both sides. The infamous Charge of the Light Brigade, when miscommunication sent British cavalry straight at Russian cannon, was in many ways the epitome of both.

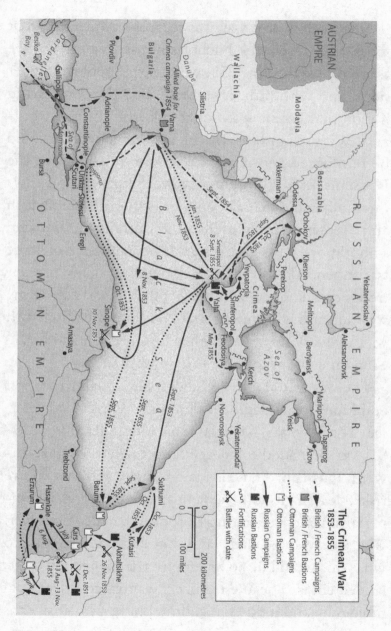

The Crimean War
1853–1855

Battles with date
Fortifications
Russian Campaigns
Russian Bastions
British / French Campaigns
British / French Bastions
Ottoman Campaigns
Ottoman Bastions

0   100 miles
0   200 kilometres

Nonetheless, this war was to prove a turning point for Russia, albeit one Nicholas himself would not see through. He died in 1855, while Sevastopol was under siege, and with him, it seemed, died the comforting notion that Russia was not at risk from its backwardness. With steamships, British and French forces could be reinforced and resupplied more quickly than the Russians, even though they were fighting on their own soil, could manage on foot. The rifles the British and French infantry used could outrange antiquated smoothbore Russian cannon. The brilliance of some tsarist generals and the stoic bravery of many of their troops could not conceal the fact that Russia's serf army was outgunned, undertrained and often badly led. It was, in every way, a metaphor for the country's social, economic and technological circumstances.

The war would prove a catalyst for arguably the most ambitious social engineering project Russia had yet to see. The new tsar, Nicholas's son Alexander II (r. 1855–81), quickly sued for peace, and turned his gaze inward. Russia had failed to modernize, and that failure risked leaving it vulnerable in an age of aggressive imperialism and a changing European balance of power. The serfs wanted their land, but had no answer as to what this would do to an empire that depended on its landed elites. The Westernizers wanted constitutional monarchy and industrialization, but had no answer as to what this would mean for Russia. The conserva-

tive Slavophiles clung to the notion that Russian culture needed to be purged of Western decadence, but
had no answer as to how that could be reconciled with
necessary modernization. Everyone agreed something
had to be done—as Tolstoy put it, "Russia must either
fall or be transformed"—but there was no agreement
as to what. All eyes turned to Alexander.

Liberation Lite

Even before his coronation, Alexander had made it
clear that he was willing to grasp the nettle Nicholas had
avoided, declaring that it was "better to abolish serfdom
from above than to wait until it begins to abolish itself
from below." Reform was necessary both to forestall
revolution but also lest foreign powers begin to feel, as
once the Poles and Swedes had, that they could meddle
in Russian politics. However, that reform would have
to be managed from above, both to ensure it was mea-
sured, but also because, between the political backwardness of the peasantry, the relative absence of a middle
class and the self-interest of the aristocracy, who else
could be trusted to do this than the tsar?

All this, though, meant that there were two fatal and
irreconcilable paradoxes to Alexander's reforms. First
of all, over liberalization. Equality before the law, constitutionalism and the like inevitably posed a challenge
to Alexander's belief that strong executive power was
needed to drive reform forward, because they are to a

large extent explicitly about limiting the untrammeled powers of state and tsar. The politically active forces that these reforms generated and required were soon forced into a choice: serve the state and lose the capacity for independent action, or be treated as subversives. The second paradox was over execution. Alexander depended on the state bureaucracy and gentry to carry out his reforms, the very people whose interests were at threat, and he could not afford to alienate them.

Nonetheless, Alexander undoubtedly earned his epithet as the "Tsar-Liberator." He pardoned political prisoners, relaxed censorship, restored the independence of the universities, established independent courts and presided over a major expansion of schools for the poor. The centerpiece of his "Great Reforms" was intended to reshape the country's agrarian and social base by finally tackling serfdom. After all, 46 million of the tsar's 60 million subjects were still serfs, and they made poor stewards of the land they did not own and indifferent soldiers for a state they did not respect. In 1861, the Emancipation Decree promised to change all that, freeing the serfs over the next two to five years, depending on their status. What they really wanted, though, was their own land, and there was the rub. Simply allowing them to take over the land they farmed would at a stroke bankrupt most of the landed gentry, so instead the serfs were forced to buy it off them, paying "Redemption Dues" over the next 49 years.

It was a classic case of a compromise that pleased no one. The serfs had for generations dreamed of the day the "little father" in St. Petersburg would finally free them, and initial enthusiasm quickly turned to anger when they realized they would have to pay—at prices they often could not afford—for the land they felt was morally theirs, earned with their blood and sweat. In 1861 alone, the army had to be called out to put down riots and protests on average more than once a day. Not that the landowners were much happier. Many had been in debt to the state, and the money they were paid for the land often simply went right back to the government. Besides, as the peasants could often not afford to keep up with their payments, even this income dried up. As if this were not enough, a major administrative reorganization that accompanied the Emancipation, with the creation of new elected local councils called *zemstvos*, meant that the rural gentry were also expected to do the state's work in the regions, from collecting taxes to dispensing justice.

Meanwhile, a new class was rising. Cities were beginning to expand, and with them a merchant class. The civil service expanded with the professionalization of many functions once handled by landlords and serf owners, and so too did the universities. Between 1860 and 1900, the number of professionally trained Russians grew from 20,000 to 85,000. Still a small fraction of the population, but for the first time this intermediary layer, neither peasant nor gentry, began to form a distinct, self-conscious intelligentsia identity. Shaped by Western ideas and Russian culture, they would form the basis for a rising revolutionary movement. In every aspect of life, new ideas were beating against the walls of the old order. Artists began to challenge the formalism of the Imperial Academy of Arts in St. Petersburg, which for over a century had imposed its stultifyingly conservative standards on the cultural scene. Wealthy women and members of the St. Petersburg literary set began agitating for more access to education, over the protests of traditionalists who, as one university rector put it, "know women's limitations better than they seem to know them themselves." From letter writing to bomb throwing, those unhappy with the existing order felt increasingly free to express their concerns and advance their agendas.

Alexander, it seems, was genuinely surprised by the response to his reforms. Demonstrating that he was his father's son, his instinctive response was to fall back on repression. A vicious circle saw heavy-handed policing and violent political protest feed off each other. In a triumph of urban romanticism, the Populists, a Russian take on the socialism then rising in the West, saw the peasant commune as some kind of utopian communist microstate. When they "went to the people" with this notion, the peasants they idealized tended to ignore them, drive them out or hand them over to the police. So they, and other groups, turned to terrorism instead, and with every prince or general they killed, the more the Gendarmerie and Third Section cracked down with the kind of vicious enthusiasm that drove more recruits into the revolutionaries' arms.

Over time, though, Alexander seems to have come to realize that intelligentsia radicalism and peasant discontent were two separate phenomena, and opted to step back from repression. On the morning of 31 March 1881, he resolved to call a commission to launch a second wave of even deeper reforms, including a constitution. And with flawless timing, the *Narodnaya Volya* ("People's Will") terrorist group, which had tried and failed to kill him seven times already, finally managed to assassinate him, that very afternoon.

The Reaction

Alexander's son and heir, Alexander III (r. 1881–94), was a narrow-minded man at the best of times and would likely have adopted a more reactionary line even had his father not just been assassinated. Under the influence of his zealot of a former tutor, Konstantin Pobedonostsev, he oppressed Russia's minority peoples—most notably the Jews—and presided over a massive expansion of repression. Police certificates of "reliability" were needed to get into university or jobs deemed "responsible," and appointed land commandants became virtual local overlords. His answer to the dilemmas of modernization was to tax the peasantry all the harsher to make money to buy what Russia needed from the West. As his finance minister Ivan Vyshnegradsky put it, "Let us starve, but let us export." (Vyshnegradsky was a multimillionaire; there was little risk of his starving, but up to half a million ordinary Russians did in the 1891–2 famine.)

When Alexander died in 1894, his son succeeded him. Nicholas II (r. 1894–1917), the last Romanov, was perhaps the most ill-starred of his line. This was a time when the Russian Empire needed a tsar with the will of Peter the Great, the wit of Catherine, the cunning of Dmitry *Donskoi*, the reformism of Alexander II and a touch of the iron conviction of Nicholas I. It got a man who was unimaginative in his conservatism, vacillating in his convictions, meek before the strong-willed, imperious to the well-meaning. He himself felt unprepared for the position. "What is going to happen to me and all of Russia?" he asked his cousin and brother-in-law, Grand Duke Alexander, on the eve of his coronation. What, indeed?

He was that most terrible of leaders, both foolish and dutiful. It quickly became clear he had no answers to the challenges facing Russia, challenges that were ironically being magnified by successful economic development. Under Vyshnegradsky's successor, Sergei Witte, the 1890s saw the economy grow some 5 percent annually. However, real average incomes actually fell, as this continued to be modernization on the cheap. The cities grew, with the worst slums being virtual no-go areas for the police and the new industrial workforce becoming susceptible to revolutionary ideas. People were hungry and angry, a combustible situation just waiting for a spark.

That spark would come from the Far East. Russia's eastward expansion had brought it into conflict over Manchuria and Korea with a rising and aggressive Japan. Believing that Japan would not fight a European power, and egged on by his cousin Kaiser Wilhelm of Germany, Nicholas put little effort into reaching a deal, and was thus stunned when the Japanese launched a surprise attack on the Russian fleet at Port Arthur in 1904. Nonetheless, he seems to have believed that, in the words of Interior Minister Vyacheslav Plehve, "a nice, victorious little war" would help reunite the country.

It soon became clear that this would be anything but. Japan had, after all, been modernizing rapidly and was fighting closer to home. At land and sea they were advancing, and so desperate were the Russians to reinforce their Pacific Squadron that they had to send the Baltic Fleet the long way there. The voyage hardly started auspiciously, when they mistook the Hull herring fishing fleet for Japanese torpedo boats and opened fire. As if that were not unimpressive enough, they only managed to sink one, and a Russian cruiser was damaged in the crossfire. War with Britain was narrowly averted, but after traveling over 18,000 nautical miles (33,000 kilometers) in seven months, the Baltic Fleet was then trounced at the Battle of Tsushima in 1905, essentially ending the war.

Maybe a victory would have helped, but an expensive and humiliating defeat undoubtedly battered the regime further. The tsar still had a certain legitimacy as God's chosen representative and the "little father" of his people. Not for long. In January 1905, more than 150,000 people staged a march to the Winter Palace in St. Petersburg to deliver a loyal petition to the tsar. They were peaceful, singing hymns and holding icons. The tsar wasn't even there at the time, but someone panicked, shooting started, and the Imperial Guards began firing volleys into the crowd. By the time the gun smoke had cleared, hundreds of protesters and bystanders were dead. And so too, arguably, were the last remnants of the tsar's standing with his people. This was the spark that was needed to ignite the country, to trigger the 1905 Revolution that Bolshevik leader Lenin would later call the "Great Dress Rehearsal" for the 1917 convulsions that would finally sweep tsarism away.

Further reading: Dominic Lieven's magisterial *Russia Against Napoleon* (Penguin, 2009) is an essential, detailed account, a worthy historical counterpart to Leo Tolstoy's epic (in every sense) *War and Peace* (there are many versions; I recommend the Anthony Briggs translation; Penguin, 2005). W. Bruce Lincoln, who writes beautifully, has the best biography of

Nicholas I (Northern Illinois University Press, 1989) and an evocative portrayal of Russia at the turn of the century, *In War's Dark Shadow* (Oxford University Press, 1983). Robert Service's *The Last of the Tsars: Nicholas II and the Russian Revolution* (Macmillan, 2017) is a judicious biography of this deeply flawed man. The Marquis de Custine's *Empire of the Czar* (Doubleday, 1989) is a contemporary account that is still very readable today.

7

"LIFE IS GETTING BETTER, COMRADES, LIFE IS GETTING BRIGHTER"

Timeline

The Lenin-Stalin Mausoleum, 1957

What does it say that an ardently secular revolutionary regime would nonetheless mummify its leader—against his own wishes—and treat Lenin in death as a saint? That the brutal dictator who murdered most of his comrades and allies then for a while joins him briefly in reverent limbo, a site for pilgrimages from the far reaches of the country? That a Communist Party preaching internationalism builds on the tsar's empire and expands its borders with equal zeal? That the Soviets, for all their professed commitment to Marxist–Leninist egalitarianism, ended up creating a near-hereditary class of Party-card-holding aristocrats every bit as rapacious and self-interested as the boyars? And that the same dilemmas—of modernization versus stability, of whether to think of themselves as Europeans or something different—continued to shape Russia's twentieth century? Perhaps a new, red flag (adopted in 1922) and murdering the last tsar and his family (by firing squad in 1918) do not quite make a wholly new country.

1905

The 1905 Revolution was not really a revolution in the sense of a coordinated effort to bring down the government. Rather, it was a wave of strikes, unrest, protests and risings, generated by a combination of frustration and anger. It did not represent an existential threat to tsarism and the status quo, but it certainly seemed that way at the time to an embattled and panicked elite. After a general strike that involved perhaps two million workers, Nicholas II issued a manifesto grudgingly promising a constitution, new freedoms of speech and religion and an elected parliament. Although the revolutionaries rejected any compromise, the more moderate forces, especially the Constitutional Democrat Party (Kadets) accepted this as a step in the right direction.

By the time this new constitution—the Fundamental Laws—was introduced in 1906, it was clear that the regime was feeling more confident. The tsar remained autocrat and the lower house of parliament, the *Duma*, was elected on a franchise weighted toward the middle and upper classes, while the upper house, the State Council, was half appointed by the tsar. The Kadets won the most seats and agitated for more sweeping constitutional change. So the government dissolved the *Duma*. A second *Duma* in 1907 saw more extreme parties win seats, including the urban Social Democrats and the rural Social Revolutionaries (both Marxist). So the government dissolved that one too, and restricted the vote more strictly to the propertied classes. The third *Duma* was suitably loyal.

As the tide of protest receded, not least following a brutal campaign of piecemeal repression, the tsar hoped to return to "normality." However, his formidable new prime minister, Pyotr Stolypin, had other ideas. Stolypin was no liberal—he had so freely used the death sentence to reimpose order that the hangman's noose became known as "Stolypin's necktie"—but a shrewd operator who realized that the system needed to restore its social foundations, although by appealing to a new constituency. The peasants had been emancipated but were still impoverished by inefficient communal land holdings. His vision was of a "wager on the strong," breaking up the communes so that the more able and industrious peasants could become a new class of *kulaks*, prosperous small landowners, the very people who were the bastion of conservatism in Germany.

"Give me twenty years of peace and you will not know Russia," he promised. But Russia didn't have 20 years of peace. The peasants resisted the end of the communes, the nobility were suspicious of any change, and Nicholas was increasingly disaffected.

When Stolypin was assassinated in 1911, in a plot possibly foreknown to the tsar, any chance at meaningful reform died with him. The revolutionary leader Lenin himself delivered arguably the most accurate epitaph: "the failure of Stolypin's policy is the failure of tsarism on this last, the last conceivable, road for tsarism."

Meanwhile, the revolutionaries were massing. The Social Democrats had split in 1906 between Lenin's Bolsheviks—the so-called "Majoritarians," although they were actually the smaller faction—and the Mensheviks. The latter saw the best chance for revolution in slowly building up a mass base of support. Lenin, instead, said that a small, disciplined body of professional revolutionaries could seize power when the time was right. They just needed an opportunity, and the Great War was about to provide them one.

War and Revolution

Karl Marx wrote that war "puts a nation to the test. As exposure to the atmosphere reduces all mummies to instant dissolution, so war passes supreme judgment upon social systems that have outlived their vitality." In this, at least, he was right, and the looming cataclysm that was the First World War was finally to end this zombie regime's existence. The balance of power in Europe had by 1914 been lost: the Austro-Hungarian, Russian and Ottoman empires were in decline, a rising

Germany was looking for its place in the sun, and colonial rivalries between Britain, France and other players were getting sharper. It was just a matter of time, and when Bosnian Serb Gavrilo Princip assassinated Archduke Franz Ferdinand of Austria on a sunny Sunday morning in Sarajevo, the dominoes started to fall. As the Austrians squared off against the Serbs, backed by the Germans, a reluctant Russia felt it had no option but to support the Orthodox Serbians. Unwilling to risk giving slow but massive Russia a chance to mobilize its forces, Germany felt it had no option but to strike first. Desperate to use the opportunity to cut a challenger down to size, France and later Britain felt they had no option but to join Russia.

The start of the war witnessed the usual grotesque spectacle of the same masses who would provide the cannon fodder greeting it with patriotic glee. Soon, though, it became clear that this, the first truly industrial modern war, posed a test Russia could not pass. By October 1917, some 15.5 million Russians had been mobilized—but more than 1.8 million were dead (and another 1.5 million civilians), 3.5 million wounded and up to 2 million captured. At peak, Russia was suffering 150,000 casualties a month, often because soldiers were being sent into battle without boots or even rifles, against machine guns and rapid-fire artillery. To feed this insatiable meat-grinder, the government was increasingly resorting to press-ganging anyone they could find. Meanwhile, the economy was near collapse. Between 1914 and 1917, prices rose by 400 percent while wages stayed static, so people were going hungry.

In a move that demonstrated a characteristic mix of imperiousness, conscientiousness and foolishness, Nicholas had made himself commander-in-chief from the first, anticipating reaping the political rewards of a quick victory. Instead, he became the embodiment of failure and hardship. That his beloved wife, Alexandra, was German-born and that he had pandered to the dissolute monk-charlatan Grigory Rasputin (until his murder in 1916) became the basis for lurid and dangerous rumors of every kind as it became clear Russia was losing the war. In February 1917, matters came to a head when the garrison of the capital, Petrograd—St. Petersburg had been renamed at the start of the war, as it sounded too German—refused to put down bread riots and even elite Guard regiments mutinied.

In Petrograd, not one but two new governments were declared. The *Duma* formed a Provisional Government, largely of Kadets, committed to founding a constitutional order. Meanwhile, revolutionaries among the workers and the soldiers looked to the Petrograd Soviet—the word simply means council—dominated by Mensheviks. Nicholas, meanwhile, was induced to abdicate by his own generals and advisers. He tried to hand power to his brother, Grand Duke Mikhail, but he recognized a poisoned chalice when presented with one. All of a sudden, there was no tsar, no representative of divine right.

As the news spread, the old order crumbled. Meanwhile, the Provisional Government and the Petrograd Soviet were locked in competition. This period is often called one of "Dual Power," but in practice it was of no power. The Kadets had all the trappings of government, but the Soviet could countermand their orders, and their reluctant commitment to continuing the war lost them much wider support. The Soviet had the streets, but was too divided and too limited to the capital to do anything with them. There was, in effect, a vacuum of power, and politics, like nature, abhors a vacuum. Someone was going to fill it, and the ruthless and pragmatic Lenin realized the moment.

That spring, the Germans—seeing him as a potential source of disruption behind enemy lines—had allowed him safe passage into Russia, and he had been building the Bolsheviks' power base. On 7 November 1917 (25 October by the old Russian calendar), they struck. However much this was later romanticized as a popular rising, mobs waving red flags in the streets, it was really an armed coup. The Bolshevik Red Guards seized the Winter Palace, the main garrisons and arsenals, and the Provisional Government melted away before it. Most of the other major cities fell to the revolutionaries too. Lenin offered the Russian people "peace, bread and land," and if many were uncertain whether he could deliver, they were at least not willing to fight to prevent him from trying. Seizing power proved to be easy. It was holding on to it that would prove hard—and the compromises the Bolsheviks would make to do so would shape the future of the Soviet regime.

Lenin vs. Lenin

In many ways, there was not one Lenin, but two. Born Vladimir Ilyich Ulyanov in 1870, he took to revolutionary politics from the age of 17 after his brother was executed for plotting to assassinate the tsar. Ruthless, indefatigable and divisive, he had spent most of his life on the run, in Siberia or in foreign exile. Like the Bolshevik party he had forged, Lenin was at once a fervent believer in an ideology whose dream was a world without oppression, misery, exploitation or want, and at the same time a merciless pragmatist, who felt that any means were justified, however bloody, if they advanced this cause.

It was Lenin-the-pragmatist who seized power in 1917. Never mind that Russia hardly seemed ready for socialism, lacking a large and politically mature working class. Never mind that, in his *The Eighteenth Brumaire of Louis Napoleon*, Marx had warned that trying to force socialism onto a country not yet prepared for it would be counterproductive, leading to a regime with conservative instincts but all the energy of revolution. (And Stalin proved him right.) Never mind all that: Lenin saw an opportunity and tied his ideology into knots to justify seizing it. After all, surely world revolution was just around the corner, and everything would work out?

Not so much. The Bolsheviks first sought to make good on their promise of peace, signing the disastrous Treaty of Brest-Litovsk that surrendered swathes of territory to the west and south—including the rich farmlands of Ukraine. However, this also unlocked the army from the frontline, and while many units simply evaporated as soldiers deserted, a collection of disgruntled generals (the so-called "Whites") began to look toward removing this usurper regime. Meanwhile, elections to a new Constituent Assembly saw the Socialist Revolutionaries, not the Bolsheviks, win a majority. Lenin had not seized power only to hand it to his rural rivals, so in January 1918 Red Guards dissolved it and the Bolshevik-dominated Congress of Soviets became the new seat of government.

More to the point, where was the bread? The cities were starving and there was no money to buy grain for them. Facing a military threat from the Whites (aided by British, US, French and even Japanese intervention), nationalist risings from various non-Russian territories, rival challenges from the Socialist Revolutionaries, and a looming collapse of the state and economy, Lenin-the-pragmatist turned to a policy called War Communism—although arguably this was more about war than communism. The democratic structures of the Soviets began to be bypassed by executive orders.

Grain was requisitioned by Red Guards at bayonet-point, and when peasants resisted, they were killed. A new secret police was founded, the Cheka, and it increasingly became a central element of Bolshevik rule.

Between 1918 and 1922, the country was racked by a vicious civil war, from which the Bolsheviks would emerge victorious, and regions such as Ukraine and the Caucasus had been reconquered, but only at a terrible cost. As many as 12 million people had died, many from famine and disease. Any traces of Lenin-the-idealist had been burned out of the Communist Party (as the Bolshevik party became known in 1918), which won by being more ruthless, disciplined and united than their numerous enemies. The Party in effect had to become the new state bureaucracy and expanded dramatically, largely by recruiting opportunists, holdovers from the old regime and politically illiterate workers. The whole culture of the Party became a war-fighting one, paranoid and savage.

After the Civil War, the state was formally named the Union of Soviet Socialist Republics (USSR), but the struggle would continue. To rebuild the economy, in 1921 Lenin instituted the New Economic Policy that liberalized some grassroots economic activity. It proved rather successful, despite intermittent crises, and meanwhile Soviet Russia also experienced an explosion of cultural, social and artistic experimentation and radical enthusiasm. This was the era of Futurist writers such as Vladimir Mayakovsky, avant-garde artists such as Kazimir Malevich, and social measures such as the

1918 "Code on Marriage, the Family and Guardian-ship" that explicitly recognized women as equal part-ners (either could choose to take the other's surname) and made divorce easy and free of blame. It was still possible to believe that, despite everything, something truly new and exciting could be built.

However, in 1922, Lenin suffered the first of a series of strokes and was largely out of the political scene. He died in 1924, but not before beginning to show seri-ous concerns about the bureaucratic police state he had created. In particular, Lenin-the-idealist was worried about the rise of Iosif Djugashvili, known as Stalin. In his testament left to the Bolshevik leadership, he made one unambiguous recommendation: "I suggest that the comrades think about a way of removing Stalin." They didn't listen to him, anxious to avoid a split in the Party, and perhaps even incredulous that Stalin, widely seen as a dull functionary—he was the general secre-tary of the Party, the job that later would mean leader of the state but at that time just the administrator-in-chief—could be such a threat. They were wrong.

Stalin quickly demonstrated his political skills, pil-ing on the eulogies of Lenin (for whom Petrograd was renamed) to mask the burial of the testament, and out-maneuvering his rivals to left and right. Compared with most of them, who were educated, cosmopolitan, Stalin also represented the rising "Civil War genera-tion" of Party officials, pragmatic, self-interested and often nationalistic to the point of racism. By the late

1920s, he was dominant, and it was clear that Stalin planned to change the country in a truly fundamental way, whatever the cost.

Terror

The age-old Russian dilemma, after all, had been how to modernize while maintaining state power. Usually, this had been attempted from above, whether Peter the Great hiring foreign shipwrights, Catherine the Great dabbling with Western philosophies or Nicholas I turning to Baltic aristocrats. Those who had tried more fundamental restructuring of the very bases of the system, such as Alexander II with emancipation or Stolypin and his "wager on the strong," had soon run into the resistance of the entrenched elites. Stalin, though, inherited a country with a new and still green elite, in a century in which the telephone and the railway, barbed wire and the machine gun, would create whole new opportunities for the dictator. He was also willing to think on a scale on which none of his predecessors had the callous coldheartedness to dream.

In 1928, he launched his "Socialism in One Country" program intended dramatically to modernize the USSR. The aim was to industrialize this sprawling state, and in the process consolidate his rule. To the twin challenges of how to afford it—where would the money come from to build factories, import Western technology and the like?—and how to impose it on a recalcitrant country, he had a single answer: terror.

The countryside was collectivized: in effect, land was nationalized and the peasantry turned into employees of the state. When the farmers resisted, they were suppressed with staggering brutality. Ukraine was brought to its knees by an engineered famine in 1932–3 that killed more than 3 million, while by 1931, at least a million peasants had been sent to the Gulag labor camps and 12 million deported to Siberia.

Collectivization was meant to bring economies of scale and new technology to farming, but more than anything else it imposed unprecedented state control over the countryside. It had been the tsarist minister Vyshnegradsky who had said, "Let us starve, but let us export," but it was Stalin who truly applied that, sending grain westward in return for money and technology, whatever the human cost at home. That also permitted industrialization, though in a clumsy and brutal way. Terror also helped motivate the workforce: real wages plummeted, but the fear of being denounced as a "wrecker" and the promise of bonuses for those "overachieving the plan" together helped keep people working. Besides, the Gulag labor camp network, which by 1939 contained 1.6 million prisoners, may have started as a place to bury the politically inconvenient, but also became a source of slave labor, from digging canals to cutting trees.

Meanwhile, Stalin turned the terror against the Communist Party itself, first staging show trials of his rivals—accusing them of being everything from spies

to saboteurs—and then systematically purging everyone who could conceivably challenge him. Culmi-
nating in an orgy of torture, mass arrests and firing
squads in 1937, this saw the Party elite broken: three-
quarters of all the representatives elected to the Party
Congress in 1934 did not survive until 1939. Even the
military high command was decimated. In 1937 alone,
90 percent of all the generals and three of the Red
Army's five marshals were purged. Art, culture, education and ideology all were turned to the glorifica-
tion of the state and Stalin; Mayakovsky, incidentally,
committed suicide in 1930, Malevich was arrested by
the secret police in the same year, and the radical social experimentation of early Bolshevism was rolled
back, in a new drive to encourage large, stable families ("We need fighters, they build this life. We need
people") and keep women in their place.

How did Stalin get away with it? He understood
power at a visceral level, and kept firm control of the
political police, in many ways the true heart of his state.
The very scale of his ambition and the consequent hor-
rors were also beyond the apprehension of most, until
it was too late. He also offered a ruthless, cannibalistic social mobility of sorts, and those willing to play
the game could hope to rise very far, very fast. For the
rest, alongside the paranoid hunt for spies and saboteurs that created its own hysteria, Stalin maintained
a huge propaganda apparatus that tapped into the very
same cultural roots as the myth of the "Good Tsar"

who was on the side of the people, only misled by his selfish advisers. "Life is getting better, comrades," he told his subjects, "life is getting brighter"—and many so desperately wanted to believe. Yet what lay around the corner was not a brave, socialist future, but the apocalypse of the Second World War.

The Great Patriotic War

In 1931, Stalin had said, "We are fifty or a hundred years behind the advanced countries. We must make good this distance in ten years. Either we do it, or they crush us." Ten years later, the Soviet Union was fighting for its life.

The USSR had been considered a pariah nation in the interwar period. As fascism rose in Europe, Stalin had first hoped to use it to reach some common cause with Britain and France, then opportunistically made his own deals with Hitler's Germany, leading to their joint partition of Poland in 1939. It is not that he did not think war with the Nazis was inevitable: he knew perfectly well Hitler saw the Soviet Union as prospective *Lebensraum*, "living space" for a new generation of Aryan master-colonists, using Slav slave labor to grow the crops and extract the resources he would need. Rather, Stalin had hoped to postpone war against Germany to give him as much time as possible to prepare.

When, in June 1941, Hitler launched Operation Barbarossa, the invasion of the Soviet Union, it came as a devastating strategic surprise. Stalin's spies, diplomats and generals had all told him what was coming, but he was sure he had Hitler's measure and that war had been delayed until next year. The Red Army was thus wholly unprepared, and by mid-July the Germans were two-thirds of the way to Moscow, most of the Soviet air force had been destroyed, and Lenin's embalmed body was being sent secretly to safety in Tyumen, 2,500 kilometers (1,500 miles) to the east. Stalin himself seems to have had something of a nervous breakdown and, for the first two weeks of the war, there was scarcely any central guidance from Moscow.

But then he recovered and threw every effort into survival. What followed were four years of phenomenal national effort, in which the invasion was gradually slowed, then halted, and finally turned around in a counteroffensive that would eventually see the Red Army crashing into Berlin, and Soviet rule imposed on Central Europe. Stalin's crude, brutal industrialization had built a war-fighting economy, and factories relocated away from the frontlines would soon be churning out the guns, planes and tanks needed. Stalin was also pragmatic: generals who had previously been sent to the Gulags as traitors were hurriedly recalled to arms, and churches that had been closed by the aggressively secular regime were reopened to enlist Orthodoxy in the struggle. The Soviets would also demonstrate once again an extraordinary will to defend the Motherland (although honesty demands that we note that this was often backed by a fear of a ruthless state). More died in the siege of Leningrad alone than the total British and American casualties of the entire war, for example.

No wonder the Russians still call this the Great Patriotic War. It is impossible to understate its importance. More than 20 million died in the war, and everybody suffered. Yet by the end, the pariah nation had become a superpower, Stalin sitting down with British Prime Minister Winston Churchill and US President Franklin D. Roosevelt at the Yalta Conference in 1945 to carve up the postwar world, with Latvia, Lithuania and Estonia directly incorporated into the USSR, and East Germany, Poland, Czechoslovakia, Hungary, Bulgaria and Romania destined to become its vassals. It seemed to have confirmed the terrible necessity of Stalin's industrialization, and the Party would be able to point to the shared experience of the war as a basis for its legitimacy.

Stalin ruled until his death in 1953, presiding over the ruthless consolidation of his puppet regimes in Central Europe and the reconstruction of the country. However, after the triumph of 1945, the limitations of Stalinism were becoming clearer. His economic model was increasingly ill-suited to the new technologies of the postwar era, and the Gulag camps were of diminishing value, not least as risings within them became more common. A restive and ambitious elite had their own agendas too. There are many indications that Stalin had decided on a new purge to cut them back down to size when he suffered a cerebral hemorrhage. He might have survived had he received prompt medical care, but he was so infamously paranoid that his guards were not allowed to check in on him and so it was too late by the time he was discovered: let no one say fate has no sense of irony.

The Long Goodbye

Stalin's successors would, in their own ways, all grapple with the familiar challenge of modernization. In the 1950s and 1960s, the Soviet Union seemed a rising power, such that when Stalin's successor, Nikita Khrushchev, told the West "we will bury you"—not quite as threatening as it sounds, but meaning that the Soviets were in the ascendant—many in the West feared the future did indeed belong to them. In hindsight, though, the real story was one of a failure of imagination and will.

Khrushchev was known for opening up the Gulags, rolling back some of the worst excesses of Stalinism and his "Secret Speech" of 1956 denouncing the old leader's "negative characteristics." In part, this was genuine, but Khrushchev had been one of Stalin's right-hand men, and he was trying to distance himself, and the Party, from the Terror. (While Stalin's body was suitably embalmed and moved into Lenin's mausoleum after his death, he was turfed out in 1961.) Khrushchev certainly had no qualms brutally crushing an anti-Soviet rising in Hungary in 1956, for example. More to the point, the Party elite began to see him as dangerous: his brinkmanship almost led to nuclear war during the 1962 Cuban Missile Crisis, and he had so mishandled the economy that there were widespread food shortages.

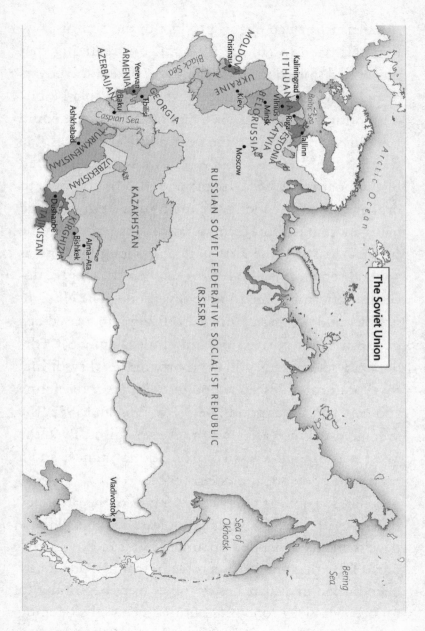

The Soviet Union

Khrushchev had been a product of the Stalinist system and tried to rule as autocrat, not realizing that power had shifted to the wider Party elite—the boyars of the new order. In 1964, he was ousted in a bloodless political coup, and his eventual successor, Leonid Brezhnev, adapted to these new political realities. He was not the *vozhd*—"boss"—but more the chairman of the board of USSR Incorporated. His role was to broker consensus among the main interest groups and bring more efficient, technocratic management to the system. As such, the first part of his lengthy time as general secretary—1964–82—seemed strikingly successful, offering something for everyone. The elite got stability and prosperity, not least through increased opportunities for corruption and embezzlement. Ordinary Soviet citizens got an improving quality of life, their political quiescence bought with laxer discipline and new consumer goods: between 1964 and 1975, the average wage increased by almost two-thirds. Even the West was offered a less confrontational stance, and a new era of detente and coexistence.

So far, so good, but the very strengths of the Brezhnev order were to be its downfall, not least because all these developments depended on ample funds to buy everyone off. By the mid-1970s, problems that had previously been buried in avalanches of rubles were beginning to surface. Massive economic ventures such as the opening up of new areas to farming had failed to deliver on their promises. A new global industrial revolution based on computing was beginning, and the USSR was falling behind. Corruption and black marketeering were eating the heart out of the official economy. A vastly expensive arms race with the West had begun. This was a slow-burning crisis that needed urgent, decisive action, but that is precisely what the aging, cautious Brezhnev couldn't and wouldn't provide. He lacked the temperament, political authority or ideas. So instead, he just survived: a metaphor for the Soviet state, becoming less capable, less healthy, more senile by the year.

By the time of his death in 1982, it was impossible to ignore the crisis. The Soviet Union was mired in a vicious conflict in Afghanistan in which boys were coming home in zinc coffins while the official media still claimed there was no war there. Poland was convulsed by nationalist protests and there were signs of restiveness in the other satellite states. The economy was stagnant, food supplies were increasingly rationed and the population was not so much rebellious as apathetic and depressed: "They pretend to pay us," went the common refrain, "and we pretend to work." Stalin's crash industrialization and the technocratic management of later years had turned peasant Russia into a Soviet nation of cities and railways, engineers and doctors, readers and writers: in 1917, only 17 percent of the population lived in cities, but this was 67 percent by 1989, and literacy rates rose from around 30 percent to near enough 100 percent. But what price such progress when your newspapers were full of lies, your leaders spoke of egalitarianism while living a privileged life such as you could not dream of enjoying, and you had to stand in line for a loaf of bread?

The new general secretary was the ascetic and acerbic Yuri Andropov, former head of the KGB political police and one of the few men largely untouched by the prevailing corruption and careerism. He was as determined to bring about change as he was unlucky: within three months of taking office he suffered kidney failure. He lasted only one more year, but his main achievement was rapidly promoting a relatively young, relatively reformist Party official called Mikhail Gorbachev. When Andropov died in 1984, Gorbachev was still not yet in a position to take over, so he adroitly backed the grayest of Party functionaries, Konstantin Chernenko, instead. After all, Chernenko was himself very ill and could be counted on to die soon. This he obligingly did in 1985, allowing Gorbachev to become general secretary. His ambition was to save the Soviet system. Instead, he was to kill it.

Reformed to Death

Gorbachev was one of the last true believers. He looked at the country, with its moribund economy, corrupt Party functionaries, demoralized workers, declining global status and threadbare Marxist–Leninist ideology, and somehow thought that it could be reformed, saved, even though he had few resources to play with and a wafer-thin majority within the ruling Politburo, or cabinet of the Party. It was a sign of unusual naivety that he could convince himself of this; it was a sign of unusual maturity that he could grow and evolve as his successive programs failed.

He began by thinking that the problem was essentially one of a few rotten apples within the *nomenklatura*, the Party elite, and a need to crack down on inefficiency and labor indiscipline. It soon became clear that it was much more systemic, and by 1986 he was talking about the need for *perestroika* (restructuring) at a more fundamental level. This meant economic modernization, but also political reform. Central to the latter was *glasnost*, usually translated as openness but really having more of a sense of speaking out. He encouraged an honest and realistic assessment of the country's problems, in part to try and convey to everyone quite how necessary reforms were, considering that for decades people had been fed a diet of comforting propaganda. After the 1986 Chernobyl nuclear disaster, when a power plant in Ukraine went into meltdown and even Gorbachev reverted to old Soviet instincts and at first tried to cover it up, he realized this had to come from below as much as above.

Increasingly, though, Gorbachev was encountering resistance and losing control. The Party bosses resented his attempts to reform. National minorities began to use new freedoms to agitate for freedom. *Glasnost* acquired a momentum of its own and all the bloody skeletons in the Party's closet began to be discovered, from corruption in the *nomenklatura* to Stalin's crimes. Instead of retreating, though, he became more radical, and in 1989 created a new constitutional basis for the country, with an elected president. Why did the general secretary of the Communist Party also need to be president? Because Gorbachev had come to realize that the Party was actually the greatest obstacle to reform and he needed an independent power base to try and force it to change.

It didn't work, though. The hardliners simply became more entrenched, and as the economy worsened, new political forces began to emerge, taking advantage of his democratization. Nationalists in Ukraine and the Baltic states began mobilizing for independence, others in Armenia and Azerbaijan began reopening old territorial disputes. Most dangerously, a former local Party boss whom Gorbachev had first promoted and then sacked, Boris Yeltsin, was rising, in due course being elected president of the Russian part of the Soviet Union. The winter of 1990–1 was a hard one, with massive miners' strikes, and Gorbachev wobbled, contemplating an alliance with the hardliners to restore order. He refused to give in to this temptation, though, and again emerged more radical than ever. He began negotiating with the elected presidents of the various constituent republics of the USSR, to agree to a new Union Treaty that would totally reshape the state, turning it from a Muscovite empire in all but name to a genuine federation of voluntary members.

This was too much for the hardliners, so in August 1991 they staged a coup, confining Gorbachev to his mansion in Crimea and declaring that an "Emergency Committee" was now in charge. They had anticipated that a cowed and docile Soviet population would simply accept their decrees. They were wrong. People began coming out onto the streets in protest, in Moscow and across the country. Had the "Emergency Committee" been as ruthless as so many previous Russian usurpers, they might have still won the day, but at that fateful moment they were not willing or able to use force. Emboldened, hundreds of protesters became thousands, and Yeltsin—whom the plotters had not even thought to have arrested—emerged as their champion.

After just three days, the coup collapsed and Gorbachev was back in Moscow, but the whole calculus of power had shifted. Yeltsin had gone along with the idea of a new Union Treaty reluctantly, largely because the risk was that otherwise the hardliners would take over. They had tried and failed, though, and Yeltsin could now indulge his deep grudge against Gorbachev. He outlawed the Communist Party and refused to sign the Union Treaty. The Baltic States declared their independence; the Ukrainians demanded theirs. Recognizing the realities of the situation, for his final duty as president of the Soviet Union, Gorbachev decreed that it would be dissolved at midnight on 31 December 1991.

The End of the Soviet Idea

The old regime was broken by the First World War, but as it had exhausted its capacity to evolve, one could argue it was already dead but just didn't yet know it. The revolutionaries who seized power in 1917 under Lenin had ruthlessness and idealism but no real blueprint for the future. The desperate struggle of the 1918–22 Civil War saw them take the country but lose their soul, idealism giving way to opportunism in a way that helped ensure Stalin's rise to power. His "Socialism in One Country" was an expression not just of his own hunger of power, but his keen awareness of the Soviet Union's vulnerabilities. He mobilized a new national myth, of the construction of socialism, in the name of a brutal campaign to modernize. Victory in the Great Patriotic War represented the apotheosis of a long-standing Russian messianism, the sense that there was something special, unique, about the country and that it had a greater destiny. In 1812, then during the revolutions of the mid-nineteenth century, they had claimed to be Europe's defenders, not its backward cousins, and now they had proof. The irony is that the savior of Europe then became the occupier of half the continent and the threat to the rest, and the

Iron Curtain not only locked Russia away from Europe, it made it more "other" than ever.

In late Soviet times, as the claim that history was on the Communist Party's side became harder and harder to believe, as corruption devoured the state from within and the economy ground slower and slower, the Kremlin was forced to rely more and more on propaganda and lies. But neither the Party nor the masses truly bought into the red-bannered fantasies that were peddled. Instead, everyone sought their own slice of Europe, from the ordinary citizens listening to the BBC in darkened rooms and swapping black-market Beatles tapes, to the elite buying themselves Scotch and imported jeans in Party-only special shops. The Soviet idea ended up as tsarism on steroids, but the Soviet people themselves had very different dreams. With the USSR over, though, were they going to be able at last to realize them?

Further reading: A good general history of the Soviet era is Robert Service's *The Penguin History of Modern Russia: From Tsarism to the Twenty-first Century* (Penguin, 2015). The best biographies of the key leaders are Service's *Lenin* (Macmillan, 2000), Simon Sebag-Montefiore's *Stalin: The Court of the Red Tsar* (Knopf, 2004), and William Taubman's *Gorbachev: His Life and Times* (Simon & Schuster, 2017). Aleksandr Solzhenitsyn's *One Day in the Life of Ivan Denisovich* (Penguin, 2000) remains the sharpest and most distilled introduction to life in the Gulag.

8

"RUSSIA HAS BEEN LIFTED BACK OFF ITS KNEES"

Timeline

Monument to The Defenders of the Russian Soil, *Moscow, 1995*

Victory Park, in Moscow's western suburbs, epitomizes how modern Russia is trying to stitch together an identity from the scraps and swathes of its history that it chooses to remember and retain. This monument, for example, unites a medieval warrior of the kind who fought with Dmitry *Donskoi* against the Mongol-Tatars at Kulikovo, with one of the infantryman who ground away at Napoleon's *Grande Armée* in 1812, and a Soviet soldier of the Great Patriotic War. Three moments of national glory, united. And why not? What country doesn't highlight its triumphs more than its miseries? The reason why it is worth dwelling a little on these three gallant defenders of the Motherland is the ideology for which they stand, which has come to define Russia in its post-Soviet days, a mix of prickly defensiveness and an inclusive nationalist myth of a unique historical mission.

The Soviets never managed to square the circle of how to "out-West the West" without surrendering the ideological myths that had become central to the Party's rule. Time and again, this held back progress, whether all the research into genetics wasted because the charlatan Trofim Lysenko managed to convince Stalin it was a "bourgeois pseudo-science," or the KGB's paranoias about the free flow of information meaning that for years photocopiers were deemed a security risk, or the way that a dogmatic insistence on central planning stifled initiative and innovation. When Gorbachev began to question the established ways, he brought the whole system down. In the process, the carefully curated—which often meant essentially falsified—recent and distant histories were suddenly open to question. Horrific details of Stalinist Terror challenged the heroic narrative of Soviet industrialization, and even the triumph in the Second World War was undermined by accounts of bad generalship and a callous disregard for the lives of soldiers. If anything, the pendulum swung too far the other way, and a tide of recovered truth, debatable opinion and outright conspiracy theory washed away any certainties. Was Lenin really a German agent? Was Stalin a pedophile? Did a UFO crash in the Russian Far East in 1986? Did Gorbachev bring down the USSR as part of a Zionist–Masonic conspiracy?

In the 1990s, Western markets and memes crashed into a Russia that was looking for new truths. Everyone bought into the idea that Russia was finally part of Europe, where it belonged. It did not take long for this assumption to become questioned; Russians who eagerly embraced Western lifestyles (when they could afford it) nonetheless began to regard themselves and their nation as being pushed down and held at arm's length by a Europe that was happy to welcome Balts and Bulgarians, Slovaks and Slovenes, but not Russians. From this came the postimperial backlash that ultimately brought Vladimir Putin to the Kremlin and in due course Russian "little green men" into Crimea and southeastern Ukraine. It also led to a new effort to construct an identity for the country, to find some meaning in its bloody, wandering journey, and from that a sense of where it should be going.

The Wild Nineties

On midnight, 31 December 1991, the Soviet Union was replaced by 15 new nations, the largest and most populous of which was the Russian Federation. But what was the new Russia? The other states had the advantage of being able to define themselves in terms of what they were not: they were no longer subjects of Moscow. The Russian Federation spanned 11 time zones; its population of some 149 million was 80 percent ethnic Russian but included minorities from Ar-

menians to Ukrainians, Tatars to Karelian Finns. Could
it lay claim to the legacies of the Soviet state? Was it a
successor to the tsarist empire?

Was this a palimpsest, dense with the half-erased texts
of past times, or a true blank sheet of paper? Now was
a time for a visionary, a leader with the passion, determination and energy to create a new Russia and bind
Russians to this dream. What they had was Boris Yeltsin.

It is easy to be dismissive of Yeltsin, especially as
through the 1990s he seemed increasingly to succumb to
drink, painkillers, health problems and self-indulgence,
presiding over a crash transition to the free market
that largely swapped state monopolies for private ones.
In a wholesale plunder of the country, entire industries were privatized, for kopeks on the ruble, into the
back pockets of selected crooks and cronies. This was
the Yeltsin who played the spoons on the head of the
president of fellow ex-Soviet state Kyrgyzstan in 1992,
who slept through a state visit to Dublin in 1994, and
whom the US Secret Service found drunk, in his un-
derwear, looking for pizza on Pennsylvania Avenue
in 1995. However, this was also the Yeltsin who had
been the face of resistance to the hardliners' 1991 coup,
who had blocked Gorbachev's efforts to reconstitute
the USSR and who, when faced with a recalcitrant
Soviet-legacy parliament in 1993, shelled it into submission. That was against the constitution, so he simply had the constitution revised retroactively to make

it legal. Then, in 1996, when it looked like the resurgent Communist Party might actually win the presidency, he turned to the oligarchs, the new mega-rich business magnates who had so profited from his free-for-all economic policies, to support him—or, as some would say, rig the election for him.

For most Russians, this was a decade of despair, uncertainty and hardship. While a handful of Russians were becoming immensely rich, most were coping with an economic crisis worse than the Great Depression of 1930s America. More than half were living below the poverty line, a bankrupt health system meant mortality skyrocketed and organized crime ran seemingly unchecked. I remember the shocking sight of lines of pensioners outside metro stations, selling anything they could find to scrape together a few rubles: an old medal, a single shoe, a half-used tube of toothpaste.

When Yeltsin had an enemy, he could be focused, ruthless, energetic; when the state-breaker had the chance to be state-maker, though, it was clear he had no real plan. There was a growing sense that this anarchy could not be allowed to continue. The country was being patronized and ignored internationally, and no wonder when it was so weak. Moscow couldn't even defeat the rebellion in its southern region of Chechnya, just fight it to an inconclusive draw. By the end of the decade, there were people in and around the Kremlin looking for a successor to Yeltsin. It had to be someone loyal and efficient (and ideally healthy and sober), someone with the determination to reestablish the power of the state, someone who could articulate some kind of vision for Russia. And they found him.

Enter the New Tsar

They settled on a relatively unknown figure, a certain Vladimir Putin. In the 1980s, he had been a KGB officer, albeit not a very distinguished one, but in the 1990s he had returned to his home city of St. Petersburg (the name Leningrad didn't survive the end of the USSR long) and in due course became deputy mayor. He began to make a name for himself as the henchman of choice, the self-effacing and efficient subordinate who had his boss's back. When mayor Anatoly Sobchak was about to be arrested on corruption charges, it was Putin who put him on a plane to France. In 1996, Putin moved to Moscow and became deputy head of the Presidential Administration's Property Management Department, where again he played a key role in keeping everything running and stamping on rumors of official embezzlement.

At this point, his career turned meteoric and, after stints as Yeltsin's deputy chief of staff and head of the Federal Security Service (successor to the KGB), on 9 August 1999, Yeltsin appointed him first deputy prime minister and, later that same day, acting prime minister. At the end of that year, Yeltsin resigned, making Putin acting president, so that he could campaign in the election that was to follow with all the advantages of incumbency. Putin repaid his debts, though: the very first decree he signed enshrined guarantees for Yeltsin, not least granting him immunity from potential corruption charges, a perk informally extended to his whole family.

Who was Putin, though? A mysterious (mysteriously convenient, some suggest) spate of apartment-building bombings across Russia in late 1999 and the eruption of a new war in Chechnya allowed him to position himself as a tough defender of security and national interests. He offered no clear program but his promise of a "dictatorship of the law" appealed to those tired of the lawlessness of the past decade. He was also, like Yeltsin in 1996, backed blatantly by state and private media alike, and duly won in the first round of the election with 53.4 percent of the vote.

He set about making it clear that the years of drift were over. The oligarchs were faced with a simple choice: accept that they no longer could dictate politics and enjoy their wealth, or pick a fight with the Kremlin and lose. Some left Russia, but the richest oil magnate, Mikhail Khodorkovsky, dared to back opposition candidates and complain about corruption. In 2003, he was arrested, charged with fraud and tax evasion and sent to prison, not least as an effective warning to the rest. The Chechens were subdued in a brutal campaign that saw their capital city, Grozny, leveled and a thuggish new local regime imposed. The Kremlin was back in business.

Putin was lucky. The Russian people were desperate for an end to the misery of the 1990s and now they had a leader who was not only sharp and energetic, he had the resources to begin to rebuild the country. The 2000s were marked by dramatic economic recovery: oil and gas accounted for almost three-quarters of Russia's exports and about half the state budget, and prices were high through the decade. Putin had the money to invest in rebuilding the country's military, to turn a blind eye to the embezzlement of his own cronies, and enough to spare for ordinary Russians, who got to enjoy an unprecedented level of comfort and security. In essence, he offered a new social contract: keep quiet and stay out of politics and I will guarantee you a steadily improving quality of life. After the shabby decay and spectacular collapse of the Soviet Union and the "Wild Nineties," this was a deal most were willing to accept.

That said, Putin would not rely on his subjects' gratitude. Russian democracy, never robust, increasingly became political theater, fake opposition parties and leaders playing their roles without hope or expectation of victory, just to keep up appearances. In Soviet times, the mass media and artists alike had been considered "engineers of the human soul," as Stalin put it: agents of the Party there to condition the masses into ideological correctness. Under Putin, instead of engineers, the media became the Kremlin's advertising executives. TV in particular (almost all eventually state-controlled or state-dominated) became the shouty, glitzy, tabloid cheerleader for his regime. In 2004, he romped home with 71 percent of the vote, and although the constitution barred him from a third consecutive term, in 2008 he simply set up his pliant prime minister, Dmitry Medvedev, as his proxy president. Putin moved from the presidential offices in the Kremlin to the prime minister's in the so-called White House, but real power moved there with him. When Medvedev's term was nearing its end, he dutifully endorsed Putin for president again and in 2012 the two men swapped offices.

Putinism on the Warpath

Meanwhile, Putin's relationship with the West had been changing. He was always an avowed Russian patriot who believed great power status was his country's birthright. At first, though, he was willing to be a partner, thinking that so long as he encouraged foreign business into Russia, and backed the USA's "Global War on Terror," then the West would treat Russia as a serious player and turn a blind eye to what went on within its borders. Soon, though, he would come to feel betrayed on both counts, and in 2007 delivered a blistering attack on Western policy in Munich, criticizing the emergence of a "unipolar"—US-dominated— world order.

Exacerbated by real and perceived Western slights and challenges, he adopted an increasingly confrontational, nationalist line. In part this was likely with an eye to his historical legacy, as the man who first saved Russia from disintegration and then, as he put it, ensured that "Russia has been lifted back off its knees." During his reign, Russia invaded neighboring Georgia (2008), annexed the Crimean Peninsula from Ukraine (2014), stirred up a civil war in Ukraine's southeastern Donbas region (2014–) and intervened in the Syrian Civil War (2015–). It also launched an aggressive campaign of intelligence operations and covert interventions, from a massive cyberattack against Estonia (2007) to the assassinations of enemies and defectors abroad.

To Putin, though, these were essentially defensive responses to Western attempts to isolate and marginalize his country and deny it global status. Support for pro-democracy and anticorruption activists in Russia, criticism of the deaths of outspoken journalists and politicians, and a series of risings against Moscow-friendly regimes in the Arab world and the post-Soviet states—especially the "Euromaidan" protests that brought down a corrupt regime in Ukraine in 2013–4—were all thrown together as evidence of a Western strategy to this effect. While the West became worried about Russian "hybrid war"—the use of subversion and disinformation to spread division and undermine political institutions—Moscow was equally concerned about facing a similar threat of its own.

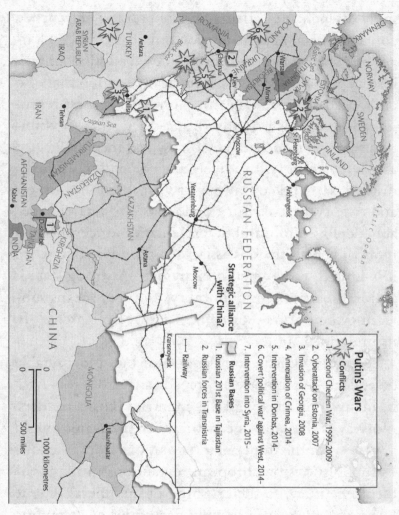

Putin's Wars

Conflicts
1. Second Chechen War, 1999–2009
2. Cyberattack on Estonia, 2007
3. Invasion of Georgia, 2008
4. Annexation of Crimea, 2014
5. Intervention in Donbas, 2014–
6. Covert 'political war' against West, 2014–
7. Intervention into Syria, 2015–

Russian Bases
1. Russian 201st Base in Tajikistan
2. Russian forces in Transnistria

Railway

Strategic alliance with China?

© HELEN STIRLING

This nationalist turn made it easier for Putin to articulate a vision for his Russia, and one that he presumably hoped would inspire a people who, since his return to the presidency in 2012, had become increas-

ingly disenchanted, tiring of fake politics, entrenched
corruption and a stagnant economy that could no longer buy their tolerance. He plundered all his country's
history freely to create a historical pedigree that also
embodied a future trajectory. One of the best repre-
sentations of this is in the *Russia—My History* exhibi-
tion that was launched in Moscow and then duplicated
across the country. In a lively and colorful multimedia
display, tsars and commissars, twelfth-century princes
and twenty-first-century diplomats, are interchange-
ably deployed in unsubtle presentation of certain pri-
mordial perspectives. Russia, for a start, is strong when
it is united, prey when it is divided. A strong state is
a moral, patriotic responsibility, and that means sub-
jecting the boyars—or commissars, or oligarchs—to
its undivided authority.

Secondly, Russia is no aggressor but a formidable
defender—the remorseless spread eastward across Asia,
its many conflicts (of all the countries Russia borders,
Norway is the only one with which it has not fought
a war—yet), the imperial interventions from Nicholas
I's suppressions of European revolutions in the nineteenth century, to the crushing of the liberal Prague
Spring in 1968, were simply necessities of defending
the Motherland and the natural order. When it pushes
back against the West, it is defending the status quo
against US-led efforts to force a "unipolar" hegemony
over the world. Even the toxic propaganda on state
TV, the growing suppression of independent watch-

dogs and the rejection of international human rights norms and monitors became spun as means of defending the Motherland against foreign interference and the "information war."

Finally, Russia is not an Asiatic country, nor yet—even though some used the term—a "Eurasian" hybrid. It is European, but proper European. It was Russians who defended Europe time and again, sometimes from enemies without, such as the Golden Horde, at others those within, whether would-be conquerors such as Napoleon or Hitler, or forces of chaos and deviance. In other words, the line is that Russia holds to the true European values at a time when the nations to its west have abandoned them. Its Orthodox faith is the genuine form of Christianity, just as its social conservatism is simply a refusal to cater to degenerate fads and postmodern moral subjectivism.

Putin and History

Of course, one could say much more about Putin. About the sometimes perversely macho public persona, about the distinctive mix of extreme (and sometimes fatal) suppression of some opposition forces and a willingness to permit and even pander to others, about whether or not, when his fourth presidential term ends in 2024, he will retire, find another constitutional workaround to retain power or pick a successor. Yet in the grand sweep of Russia's extraordinary history, ought he not be treated as just another tsar or general secretary, deserving of a section or two, but no more? To be sure, he deserves full credit for stabilizing the country at home and restoring to it a role, an antagonistic and sometimes petulant one, on the world stage. Yet he has not been as murderous as Ivan (the Terrible) or Stalin (the much more terrible), not as (quite literally) larger than life as Peter the Great. He lacks the coldly ruthless intellect of a Lenin or an Andropov, or the delicate political instincts of a Catherine the Great or Dmitry *Donskoi*.

That is not to minimize Putin, but to put him in his place. He has certainly tried to shape Russia's understanding of its history. Increasingly, school textbooks and university courses must cleave to an official version that maximizes the triumphs, and minimizes the tragedies. Stalin becomes a necessary modernizer and a war leader, with the Gulags relegated to the margins. Putin demanded that this new, official story of the country should be something "free of internal contradictions and double interpretation"—as if true history was ever so neat.

He's not the first to try to dictate Russia's image and historical record. Dmitry *Donskoi* had his tame chroniclers, Catherine the Great carefully curated her country's profile in Europe, and the cult of "Official Nationality" under Alexander III was accompanied by a campaign to muzzle and housetrain pesky scholars who insisted on challenging its precepts. Most strikingly of all, the official *History of the All-Union Com-*

munist Party (Bolsheviks): Short Course, edited by Stalin and released in 1938, was an attempt to reframe events even in living memory. In the next 20 years, more than 42 million copies of the *Short Course* were printed and distributed, in 67 languages, possibly making it the most widely read book after the Bible.

The point is that none of these attempts worked, not in their true intent of being able to shape the Russians' understanding of themselves. A palimpsest people and a country with no sharp geographic, cultural or ethnic borders may be all the more eager for national myths that help unite and define them, but they are also that much harder to confine in any one story, "free of internal contradictions and double interpretation."

Putin absolutely fits within the wider patterns of Russian history, although probably as an essentially transitional figure, neither Soviet nor truly post-Soviet. The USSR was clearly falling behind the West, unable to compete in a new arms race, and its international position was thus increasingly vulnerable. Gorbachev ushered in an attempt to modernize the Soviet Union, which necessarily involved liberalization, and this brought unrest and eventual collapse. To Putin, this was "a major geopolitical catastrophe of the century"—though to be fair, that does not mean he wants to restore the USSR—and it reflected weakness on the part of the government. After the new "Time of Troubles" of the Yeltsin era, Putin has come to see the greater threat coming from domestic weakness—possibly supported by hostile foreign powers—and thus, for all the investment in drones for the military and rockets for orbit, as well as his adventurism abroad, his regime is essentially conservative. He is Nicholas I, holding the line against disorder; Patriarch Nikon, restoring older orthodoxies; perhaps at most Peter the Great, happy to adopt technologies from the West to arm the state and control the elite, but not to reform from below.

Hypertext Palimpsest and Its Ironies

Meanwhile, the palimpsest gains more and more layers of superimposed script. Putin's own generation, of the *Homo sovieticus* who was not only born and raised in Soviet times but also had a formative early career years before 1991, is dominant, but it is being challenged by new generations, some shaped by the wild 1990s, some who have not even known a Russia as adults in which Putin wasn't in charge. There are those who rebel, joining a beleaguered but vibrant civil society, looking West for inspiration and ambitions. Others blend the orthodoxies of Putinism with a hipster cynicism, embracing Russia's new global status as the international bad boy and putting it on a T-shirt. "Putin: the most polite of people," reads one, riffing off the Russian name for what the West called the "little green men," the commandos who seized Crimea. "Isolate us? Yes please!" reads another, along with a McDonald's logo, LGBT symbol and protest placard, all crossed out with red X's.

At the same time, things are getting more complex, not less. There is a new, huge mosque in Moscow, near the Olympic stadium, as Muslims from the North Caucasus and Central Asia arrive as both citizens and—especially the latter—temporary guest workers. With them come new influences, from Caucasian restaurants to the vertical Afghan bazaar that has largely taken over the Soviet-era Hotel Sevastopol. Putin had a huge statue of St. Vladimir—Grand Prince Vladimir the Great—erected outside the Kremlin, but he was Vladimir of Kiev, and just as Kiev is now Kyiv, Ukraine is not just an independent country, it is one increasingly looking west, not east. Is Vladimir still Russia's cultural property? Or is he truly Ukraine's Volodymyr now? In Moscow's airports, there are now special passport lanes for Chinese package tourists, and more and more signs are in Chinese as well as English. In the Russian Far East, a flood of Chinese money is reshaping whole cities and regional economies. As one Russian scholar told me of his students, "They learn English because of the heart, Chinese because of the head.".

Nor are all the influences played out in the physical geographies of Russia. The palimpsest is acquiring hypertext, links into cyberspace in which information and cultural influence flow freely back and forth. Three-quarters of Russians regularly use the internet, and they use it as much as the average American. Many get their news online from foreign sources, watch videos from abroad and, as importantly, form online communities that cross borders. From discussion boards to gaming clans, Russians are not just trolls and troublemakers, they are actively engaging in new, virtual fellowships and movements.

The irony is that by defining "his" Russia in many ways in opposition to Europe and the West—challenging everything from its international order to its social values—Putin is, like so many Russian leaders before him, letting the outside world define him and the country. That's a very commonplace characteristic, though, true of almost every Russian ruler since Ivan *Grozny* brought Russia into Nordic politics and offered England's "Virgin Queen" Elizabeth I his bloody hand in marriage.

The greater irony is in Putin's efforts actively to mobilize myths of every kind in support of Russian exceptionalism, the notion that its history grants it a special and heroic role in the world. In this, he is tapping everything from the status of Moscow as the "Third Rome" to Kulikovo. Yet the very effort the Kremlin's "political technologists" and compliant historians have to put in to trying to persuade Russians that they are a special people, apart from Europe and embattled by its malign cultural and geopolitical forces, demonstrates that they are swimming against the tide.

After all, even the Russians who still revere Putin and quite literally wear the T-shirt eagerly learn English, de-

vour Western films and TV programs, and seek also in
their own cultural creations to fit into this mainstream.
We should remember that this is a country in which
on one side of a street one can see a huge mural dominating a whole facade of a tower block, exalting some
great Russian general, and on the other—a surreal experience I had myself—an equally huge mural advertising the release of a Hollywood blockbuster; and not just
any blockbuster, but *Captain America*. Much is made of
the transformation of Moscow into a lively, vibrant and
beautiful city, but just as St. Petersburg was designed by
Europeans, a great deal of this is thanks to Western architects. From the Dutch team remodeling Tverskaya,
the city's premier avenue, to the American DS+R design
bureau, creators of New York's wonderful High Line
public park, which defined a huge new green space at
Zaryadye, right next to Red Square, the Russian capital
is being rebuilt by Westerners as a European city.

Thanks to shared historical experiences and growing cross-national trade, to the internet and Hollywood
blockbusters, to cheap package holidays in Spain and
Cyprus, and mutual concerns about the rise of China,
Russia really is closer to Europe than at any point in its
history. Technically, Europe ends at the Ural Mountains,
halfway across Russia, but the Europe of the mind rolls
all the way to Vladivostok on the Pacific. When sur-
veyed, most Russians agreed with the statement "Russians are Europeans"—but some of the highest figures

for this came precisely from the distant east of the country, where "Asian" is not just an abstract concept but an immediate reality.

This is a country with a rich heritage and still vast untapped human potential. It is all too easy to see today's Russia simply in terms of TV news video grabs: the warplanes over Syria, the riot police in the streets, the self-satisfied fat cats on their top-of-the-range yachts, and Putin, the solitary figurehead of this again-menacing nation. Yet there is vastly more to it than that. Of course, there is the rich cultural legacy, of Tolstoy and Tchaikovsky, *Battleship Potemkin* and the Bolshoi, just as its history, for all its goriness, has more than its share of triumphs, heroism and generosities glittering in the darkness. But in many ways, these are the raw materials of the old Russian tradition of seeking the future from the past.

Besides, how many times, after all, can a palimpsest be written over, erased and emended, before you simply have to start with a new page? To quote Marx one last time, "Traditions of dead generations weigh like a nightmare on the minds of the living." (Writing as a non-Marxist, it is striking how Russia-relevant so much of his gloomy pronouncements turn out to be.) When does one wake up from the nightmare and move on? This is a country that is far more than the sum of its historical achievements. A new generation of activists and entrepreneurs, scientists and artists, thinkers and dreamers, are consciously trying to find new paths for Russia, not just choosing which old one to walk again. More substantially, when Russians are polled about what they want for their future, their country's great power status and fear for its security come far down the list. Instead, they crave not just a decent life, but freedoms to speak, organize and protest, an end to corruption and a chance to feel they can have some meaningful impact on how their society is organized—all the freedoms we take for granted in the West. Perhaps, after centuries torn between a desperate desire to be accepted by the rest of Europe and a defiant determination to stand alone, Russia has a chance simply to be itself. After all, the irony of "Europe" is that the centripetal pressures brought about by the European Union, its expansion east and south, and Brexit, all demand a growing awareness that there is no one "Europe." There is the Europe of Sweden and Germany, but also the Europe of Italy and Greece, that of Hungary, that of the Balkans, and that of the UK. There is room for Russia, if the Russians are feeling willing to come to terms with themselves. Putin and his cohorts may try to persuade themselves—and their people— otherwise, but the notion that they are not becoming more European is the final myth of all.

Further reading: Chrystia Freeland's *Sale of the Century: The Inside Story of the Second Russian Revolution* (Abacus, 2005) tells the story of the 1990s well through the financial shenanigans of the time. The best takes on Putin are Fiona Hill and Clifford G. Gaddy, *Mr. Putin: Operative in the Kremlin* (Brookings, 2015), on the man, and Anna Arutunyan, *The Putin Mystique* (Skyscraper, 2014), on the country that bred him. For my own take, *We Need to Talk about Putin* (Ebury, 2019) distils my thinking about the man into a slim volume. Mikhail Zygar's *All the Kremlin's Men: Inside the Court of Vladimir Putin* (PublicAffairs, 2016) is a brilliant look at all the other people around the new tsar.

ACKNOWLEDGMENTS

A book of such a historic sweep is a challenging but also exciting project, distilling 12 busy centuries into relatively few pages. As such, I have made all kinds of simplifications and omissions, and also have inevitably depended on the ideas and inspirations of many of my colleagues, sometimes directly but as often simply through intellectual osmosis, as their words and ideas have informed my own. Some are name-checked in the "Further reading" sections through the book, but some deserve specific note. Dominic—Chai—Lieven was an inspirational supervisor to my PhD and a thoughtful and generous senior colleague since, whose views on tsarism and empires in general have helped enrich my thinking throughout my career. I am indebted to Peter Jackson for comradeship and co-teaching at Keele University, and much of my thoughts on the early centuries of the

Rus' and the Golden Horde are derived from his scholarship and my no-doubt-imperfect recollections of his words. Finally, I never met W. Bruce Lincoln and, as he died in 2000, never will, but I also want to note how his writings inspired me with their self-evident proof that it is possible to write good history that is also engaging prose. There are many more who could and should have been mentioned, and my humble apologies to all those I have slighted in what otherwise could have been an acknowledgments section the length of a chapter.

On a more personal level, I also want to thank those who read earlier drafts and gave their comments: Anna Arutunyan, Daria Mosolova, Robert Otto and Katherine Wilkins. I appreciate their generosity with their time and thoughts, and apologize for any remaining errors and infelicities, which are all my own.

Likewise, my thanks to Robyn Drury at Ebury Publishing and Peter Joseph and Grace Towery at Hanover Square Press for their enthusiastic support and to the copy editor, Howard Watson, for his meticulous and sympathetic work.

Finally, I do want to record my appreciation of the European University Institute's Robert Schuman Centre for the invitation to be a Jean Montnet Fellow there in 2018–19, and its director, Brigid Laffin, for her support. The first outlines of this book were hashed out in the EUI's Tuscan hillside fastness: there are perks to an academic's life.

INDEX

Page numbers in *italics* indicate illustrations.